The Oxford Book of Marriage

HELGE RUBINSTEIN has been married for 35 years and has four children and two grandchildren. After taking a degree in Modern Languages at Somerville College, Oxford she began writing cookery books while bringing up her family. These include *The Penguin Freezer Cookbook* (1973), *Ices Galore* (1977) and *The Chocolate Book* (1981). She has worked as a marriage guidance counsellor for nearly twenty years, and was Chairman of the London Marriage Guidance Council for five years.

The Oxford Book of Marriage

EDITED BY
HELGE RUBINSTEIN

Oxford New York
OXFORD UNIVERSITY PRESS
1992

Oxford University Press, Walton Street, Oxford OX2 6DP
Oxford New York Toronto
Delhi Bombay Calcutta Madras Karachi
Petaling Jaya Singapore Hong Kong Tokyo
Nairobi Dar es Salaam Cape Town
Melbourne Auckland
and associated companies in
Berlin Ibadan

Oxford is a trade mark of Oxford University Press

First published 1990 by Oxford University Press
First issued as an Oxford University Press paperback 1992

British Library Cataloguing in Publication Data
Data available
ISBN 0-19-282930-0

Library of Congress Cataloging in Publication Data
The Oxford book of marriage/edited by Helge Rubinstein.
p. cm. Includes index.
1. Marriage—Literary collections. I. Rubinstein, Helge.
808.8'0354—dc20 [PN6071.M2094 1992] 91-28857
ISBN 0-19-282930-0

Printed in Great Britain by
Clays Ltd.
Bungay, Suffolk

FOR H

ACKNOWLEDGEMENTS

ALMOST everyone, it seems, has a favourite passage about marriage. I have received generous suggestions for this book from many people, strangers as well as friends, and I am very grateful to them all.

I would like to thank particularly: Godfrey Smith, who trawled the readers of his column in the *Sunday Times*, especially for material in the area where I had found a dearth—namely accounts of happy marital sex; Caroline Raphael, whose suggestions were always sensitive and rewarding; Bernard McGinley who led me in many interesting directions; Steve Barfield and Pier Bryden, who helped by tracing half-remembered passages, as well as making useful suggestions of their own.

Above all, I am grateful to my husband Hilary, who encouraged me to see wood when I was overwhelmed by trees—and who makes the subject of the anthology such a pleasure.

H.R.

CONTENTS

INTRODUCTION

EVERY marriage is unique—the happy (*pace* Tolstoy) as well as the unhappy—nor is any marriage wholly the one or the other. But there are some stages and turning-points in the life of a marriage, as there are in the development of an individual, that are common to most couples. The purpose of this anthology is to illuminate these through the insights of poets, novelists, essayists, playwrights, and men and women who, in their letters and diaries, have described with particular freshness and honesty their own feelings and experiences. The best writing on marriage shows us a landscape we recognize and helps us to understand our own lives and reactions.

The book aims to present a panorama of this inner landscape, and can be read as a kind of life-story of marriage, as well as dipped into at random. Each chapter is devoted to one aspect or stage of marriage. Many have a central passage, often quite lengthy, which captures a crisis or encapsulates an essential experience belonging to that phase. Some sections have a preponderance of poetry—there are critical moments or intense emotions that are best distilled in poetic form. Others, notably those that deal with the ongoing processes of a relationship or the more domestic aspects of marriage, are more suitably described in prose.

I have drawn on sources from different languages and over a span of two thousand years. Inevitably, there are some writers, those who have written with exceptional insight or particular subtlety about the nuances of a relationship, and whose relevance today is unaffected by social change or literary conventions, whose writings are quoted more often. I have found particularly rich material in fiction from the nineteenth and twentieth centuries, as novelists have become increasingly introspective.

There are a number of celebrated passages about marriage which I have omitted because they do not fit the framework of this book. These include the innumerable bits of advice given to young men and women on entering matrimony which no longer convey any relevant wisdom to our generation. Nor have I included the countless marital aphorisms, epigrams, wisecracks, and old saws, most of which scoff at the institution or at the poor fools who put their heads into its noose, and which offer at best only half-truths.

It is easy to ridicule marriage, and many believe that it is on trial in the second half of the twentieth century, with scant support from state or society, and little even from the church. Yet it has survived every social upheaval so far, and there is no sign that couples are chary of entering into marriage contracts, even if the bond is no longer always consecrated in conventional ways, nor necessarily seen as a lifelong commitment.

The feminist revolution of this century has provided the most powerful challenge to traditional patterns of marriage. Yet, paradoxically, it may also have strengthened the institution by giving greater freedom to both partners, and by allowing men to accept some of the traditionally female values, which rate success in personal relationships as highly as conventional worldly success. While it has been difficult to find appropriate passages about marriage in contemporary feminist writing, I have been struck by the many feminist sentiments to be found in books by eighteenth- and nineteenth-century writers, men as well as women.

Having worked for nearly two decades as a marriage counsellor (and been married for more than three), I am constantly fascinated by the complex dynamics of marriage. There is no universal recipe for success. Though we know that the emotional luggage that each partner brings to the union will affect its course, we still have no reliable explanation why for some marriage becomes a stifling prison or a spirit-crushing battlefield, while others are released by it from the hurts and disappointments of their early life.

At its best, marriage becomes a crucible for psychological growth, allowing the individual to break through inhibitions and self-imposed limitations, and provides the most nurturing, healing, and fulfilling area of a person's life. Certainly, no better institution has yet been found that satisfies the fundamental human need for intimate communion with another. Though this book deals as much with its darker aspects, I hope that ultimately it will stand as a celebration of married love and the institution of marriage.

The Oxford Book of Marriage

I

Decisions, Choices, and Recognition

MARRIAGE has had its protagonists and antagonists through the ages, and the debate is by no means over. Even among today's most ardent feminists, not all would wish to abolish marriage completely, and the institution seems flexible enough to adapt itself to constantly changing social conditions.

For the individual, to marry or not to marry may sound like a rational question, but reason rarely has much to do with the answer.

Both Darwin and Kafka weigh up the pros and cons, but in the end they base their decisions on emotional, not intellectual grounds. As Pantagruel chides his pupil more than three centuries earlier, 'are you not assured within yourself what you have a mind to? The chief and whole point of the matter lieth there.' Such momentous life-decisions should only emerge from internal conviction.

And we generally choose our partners in the same way. 'I'll know when my love comes along' is the popular song's shorthand for the radiant experience of falling in love, a recognition compounded of all our previous history, conscious and unconscious, of our need to relive certain early experiences or to rewrite our emotional life-story, to heal old wounds, and to fulfil unacknowledged needs. Garibaldi, writing his memoirs years after the event, describes his first meeting in Brazil with the woman he carried off secretly in his ship, and who stayed by his side through jungle and battles: 'We both remained enraptured and silent, gazing on one another like two people who meet for the first time, and seek in each other's faces something which makes it easier to recall the forgotten past.' As Shaw would say, it is the Life Force at work.

So ancient is the desire of one another which is implanted in us, reuniting our original nature, seeking to make one of two, and to heal the state of man. Each of us when separated, having one side only, like a flat fish, is but the tally-half of a man, and he is always looking for his other half. . . .

Suppose Hephaestus, with his instruments, to come to the pair who are lying side by side and to say to them, 'What do you mortals want of one another?' they would be unable to explain. And suppose further, that when he saw their perplexity he said: 'Do you desire to be wholly one; always day and night in one another's company? for if this is what you desire, I am ready to melt and fuse you together, so that being two you shall become one, and while you live live a common life as if you were a single man, and after your death in the world below still be one departed soul, instead of two—I ask whether this is what you lovingly desire and whether you are satisfied to attain this?'—there is not a man of them who when he heard the proposal would deny or would not acknowledge that this meeting and melting into one another, this becoming one instead of two, was the very expression of his ancient need. And the reason is that human nature was originally one and we were a whole, and the desire and pursuit of the whole is called love.

PLATO, *The Symposium* (trans. Benjamin Jowett), *c.*400 BC

'If there were no authority on earth
Except experience, mine, for what it's worth,
And that's enough for me, all goes to show
That marriage is a misery and a woe;
For let me say, if I may make so bold,
My lords, since when I was but twelve years old,
Thanks be to God Eternal evermore,
Five husbands have I had at the church door;
Yes, it's a fact that I have had so many,
All worthy in their way, as good as any.
'Someone said recently for my persuasion
That as Christ only went on one occasion

To grace a wedding—in Cana of Galilee—
He taught me by example there to see
That it is wrong to marry more than once.
Consider, too, how sharply, for the nonce,
He spoke, rebuking the Samaritan
Beside the well, Christ Jesus, God and man.
'Thou hast had five men husband unto thee
And he that even now thou hast,' said He,
'Is not thy husband.' Such the words that fell;
But what He meant thereby I cannot tell.
Why was her fifth—explain it if you can—
No lawful spouse to the Samaritan?
How many might have had her, then, to wife?
I've never heard an answer all my life
To give the number final definition.
People may guess or frame a supposition,
But I can say for certain, it's no lie,
God bade us all to wax and multiply.
That kindly text I well can understand.
Is not my husband under God's command
To leave his father and mother and take me?
No word of what the number was to be,
Then why not marry two or even eight?
And why speak evil of the married state?
'Take wise King Solomon of long ago;
We hear he had a thousand wives or so.
And would to God it were allowed to me
To be refreshed, aye, half so much as he!
He must have had a gift of God for wives,
No one to match him in a world of lives!
This noble king, one may as well admit,
On the first night threw many a merry fit
With each of them, he was so much alive.
Blessed be God that I have wedded five!
Welcome the sixth, whenever he appears.
I can't keep continent for years and years.
No sooner than one husband's dead and gone
Some other Christian man shall take me on,
For then, so says the Apostle, I am free
To wed, o' God's name, where it pleases me.

Wedding's no sin, so far as I can learn.
Better it is to marry than to burn.

*

Virginity is indeed a great perfection,
And married continence, for God's dilection,
But Christ, who of perfection is the well,
Bade not that everyone should go and sell
All that he had and give it to the poor
To follow in His footsteps, that is sure.
He spoke to those that would live perfectly,
And by your leave, my lords, that's not for me.
I will bestow the flower of life, the honey,
Upon the acts and fruit of matrimony.

GEOFFREY CHAUCER, *The Wife of Bath's Tale: Prologue*, c.1390

But what do I trouble myself to find arguments to persuade to, or commend marriage? behold a brief extract of all that which I have said, . . .

1. Hast thou means? thou hast one to keep and increase it.—2. Hast none? thou hast one to help to get it.—3. Art in prosperity? thine happiness is doubled.—4. Art in adversity? she'll comfort, assist, bear a part of thy burden to make it more tolerable.—5. Art at home? she'll drive away melancholy.—6. Art abroad? she looks after thee going from home, wishes for thee in thine absence, and joyfully welcomes thy return.—7. There's nothing delightsome without society, no society so sweet as matrimony.—8. The band of conjugal love is adamantine.—9. The sweet company of kinsmen increaseth, the number of parents is doubled, of brothers, sisters, nephews.—10. Thou art made a father by a fair and happy issue.—11. Moses curseth the barrenness of matrimony, how much more a single life?—12. If nature escape not punishment, surely thy will shall not avoid it.

All this is true, say you, and who knows it not? but how easy a matter is it to answer these motives, and to make an *Antiparodia* quite opposite unto it? To exercise myself I will essay:

1. Hast thou means? thou hast one to spend it.—2. Hast none? thy beggary is increased.—3. Art in prosperity? thy happiness is ended.—4. Art in adversity? like Job's wife she'll aggravate thy misery, vex thy soul, make thy burden intolerable.—5. Art at home?

she'll scold thee out of doors.—6. Art abroad? If thou be wise keep thee so, she'll perhaps graft horns in thine absence, scowl on thee coming home.—7. Nothing gives more content than solitariness, no solitariness like this of a single life.—8. The band of marriage is adamantine, no hope of loosing it, thou art undone.—9. Thy number increaseth, thou shalt be devoured by thy wife's friends.—10. Thou art made a cornuto by an unchaste wife, and shalt bring up other folks' children, instead of thine own.—11. Paul commends marriage, yet he prefers a single life.—12. Is marriage honourable? What an immortal crown belongs to virginity?

*

'Tis a hazard both ways I confess, to live single or to marry, *Nam et uxorem ducere, et non ducere malum est*, it may be bad, it may be good, as it is a cross and calamity on the one side, so 'tis a sweet delight, an incomparable happiness, a blessed estate, a most unspeakable benefit, a sole content, on the other, 'tis all in the proof. Be not then so wayward, so covetous, so distrustful, so curious and nice, but let's all marry.

ROBERT BURTON, *The Anatomy of Melancholy*, 1621

How Panurge asketh Counsel of Pantagruel whether he should marry, Yea or No.

Panurge . . . fetching, as far from the bottom of his Heart, a very deep sigh, said, My Lord and Master, you have heard the Design I am upon, which is to marry, if by some disastrous mischance, all the Holes in the World be not shut up, stopped, closed, and bush'd. . . .

Nevertheless, . . . if I understood aright that it were much better for me to remain a Batchelor as I am, than to run headlong upon new hair-brain'd Undertakings of Conjugal Adventure, I would rather choose not to marry, quoth *Pantagruel*. Then do not marry. Yea, but (quoth *Panurge*) would you have me so solitarily drive out the whole Course of my Life without the Comfort of a Matrimonial Consort? You know it is written, *Væ soli*, and a single Person is never seen to reap the Joy and Solace that is found with married Folks. Then marry, in the Name of God, quoth *Pantagruel*. But if (quoth *Panurge*) my Wife should make me a Cuckold; as it is not unknown unto you, how this hath been a very plentiful Year in the production of that kind of Cattel; I would fly out, and grow impatient, beyond all measure and mean. I love Cuckolds with my Heart, for they seem unto me to be of a right honest Conversation, and I, truly, do very willingly frequent their Company: but should I die for it, I would not be one of their number, that is a Point for me of a two-sore prickling Point. Then do not marry (quoth *Pantagruel*) for without all controversie, this Sentence of *Seneca* is infallibly true, *What thou to others shalt have done, others will do the like to thee.* Do you (quoth *Panurge*) aver that without all exceptions? Yes, truly, (quoth *Pantagruel*) without all exception. Ho, ho (says *Panurge*) by the Wrath of a little Devil, his meaning is, either *in this world*, or in *the other*, which *is to come*. Yet seeing I can no more want a Wife, then a blind Man his Staff, the Funnel must be in agitation, without which manner of Occupation I cannot live, were it not a great deal better for me to apply and associate my self to some one honest, lovely, and vertuous Woman, then (as I do) by a new change of Females every Day, run a hazard of being Bastinadoed, or (which is worse) of the Great Pox, if not of both together: For never (be it spoken, by their Husbands leave and favour) had I enjoyment yet of an honest Woman. Marry then in God's Name, quoth *Pantagruel*. But if (quoth *Panurge*) it were the Will of God, and that my Destiny did unluckily lead me to marry an honest Woman who should beat me, I would be stor'd with more than two third parts of the Patience of *Job*, if I were not stark mad by it, and quite distracted with such rugged Dealings: for it hath been told me, that those exceeding honest Women have ordinarily very wicked Head-pieces; therefore it is that their Family lacketh not for good Vinegar. Yet in that case should it go worse with me, if I did not then in such sort bang her Back and Breast, so thumpingly bethwack her Giblets, to wit, her Arms, Legs, Head, Lights, Liver, and Milt, with her other Intrails, and mangle, jag, and slash her Coats, so after the Cross billet fashion, that the greatest Devil of Hell should wait at the Gate for the reception of her damned Soul. I could make a shift for this Year to wave such molestation and disquiet, and be content to lay aside that trouble, and not to be engaged in it.

Do not marry then, answered *Pantagruel*. Yea, but (quoth *Panurge*) considering the Condition wherein I now am, out of Debt and Unmarried; mark what I say, free from all Debt, in an ill hour, (for were I deeply on the Score, my Creditors would be but too careful of my Paternity) but being quit, and not married, no Body will be so regardful of me, or carry towards me a Love like that which is said to be in a Conjugal Affection. And if by some mishap I should fall sick, I would be lookt to very waywardly. . . .

Marry then in the Name of God, quoth *Pantagruel.*

*

Your Counsel (quoth *Panurge*) under your Correction and Favour, seemeth unto me not unlike to the Song of Gammer *Yeabynay*; it is full of Sarcasms, Mockqueries, bitter Taunts, nipping Bobs, derisive Quips, biting Jerks, and contradictory Iterations, the one part destroying the other. I know not (quoth *Pantagruel*) which of all my Answers to lay hold on; for your Proposals are so full of *ifs* and *buts*, that I can ground nothing on them, nor pitch upon any solid and positive Determination satisfactory to what is demanded by them. Are not you assured within your self of what you have a mind to? the chief and main point of the whole matter lieth there; all the rest is merely casual, and totally dependeth upon the fatal Disposition of the Heavens.

We see some so happy in the fortune of this Nuptial Encounter, that their Family shineth (as it were) with the radiant Effulgency of an Idea, Model or Representation of the Joys of Paradice; and perceive others again to be so unluckily match'd in the Conjugal Yoak, that those very basest of Devils, which tempt the Hermits that inhabit the Deserts of *Thebaida* and *Montserrat*, are not more miserable than they. It is therefore expedient, seeing you are resolved for once to take a trial of the state of Marriage, that, with shut Eyes, bowing your Head, and kissing the Ground, you put the business to a Venture, and give it a fair hazard in recommending the success of the residue to the disposure of Almighty God. It lieth not in my Power to give you any other manner of Assurance, or otherways to certifie you of what shall ensue on this your Undertaking.

FRANÇOIS RABELAIS, *Gargantua and Pantagruel* (trans. Sir Thomas Urquhart), 1532–52

Pencilled notes written by Charles Darwin in 1837 or 1838:

THIS IS THE QUESTION

MARRY

Children—(if it please God)—constant companion, (friend in old age) who will feel interested in one, object to be beloved and played with—better than a dog anyhow—Home, and someone to take care of house—Charms of music and female chit-chat. These things good for one's health. Forced to visit and receive relations *but terrible loss of time*.

My God, it is intolerable to think of spending one's whole life, like a neuter bee, working, working and nothing after all.—No, no won't do.—
Imagine living all one's day solitarily in smoky dirty London House.—Only picture to yourself a nice soft wife on a sofa with good fire, and books and music perhaps—compare this vision with the dingy reality of Grt Marlboro' St. Marry—Marry—Marry Q.E.D.

Not MARRY

No children, (no second life) no one to care for one in old age.—What is the use of working without sympathy from near and dear friends—who are near and dear friends to the old except relatives.
Freedom to go where one liked—Choice of Society *and little of it*. Conversation of clever men at clubs.—
Not forced to visit relatives, and to bend in every trifle—to have the expense and anxiety of children—perhaps quarrelling.
Loss of time—cannot read in the evenings—fatness and idleness—anxiety and responsibility—less money for books etc.—if many children forced to gain one's bread.—(But then it is very bad for one's health to work too much)
Perhaps my wife won't like London; then the sentence is banishment and degradation with indolent idle fool—

On the reverse side he wrote:

It being proved necessary to marry—When? Soon or Late. The Governor says soon for otherwise bad if one has children—one's character is more flexible—one's feelings more lively, and if one does not marry soon, one misses so much good pure happiness.—

But then if I married tomorrow: there would be an infinity of trouble and expense in getting and furnishing a house,—fighting about no Society—morning calls—awkwardness—loss of time every day—(without one's wife was an angel and made one keep industrious)—Then how should I manage all my business if I were obliged to go every day walking with my wife.—Eheu!! I never should know French,—or see the Continent,—or go to America, or go up in a Balloon, or take solitary trip in Wales—poor slave, you will be worse than a negro—And then horrid poverty (without one's wife was better than an angel and had money)—Never mind my boy—Cheer up—One cannot live this solitary life, with groggy old age, friendless and cold and childless staring one in one's face, already beginning to wrinkle. Never mind, trust to chance—keep a sharp look out.—There is many a happy slave—

¶ *He married Emma Wedgwood on 29 January 1839.*

21 July 1913

Summary of all the arguments for and against my marriage:

1. Inability to endure life alone, which does not imply inability to live, quite the contrary, it is even improbable that I know how to live with anyone, but I am incapable, alone, of bearing the assault of my own life, the demands of my own person, the attacks of time and old age, the vague pressure of the desire to write, sleeplessness, the nearness of insanity—I cannot bear all this alone. I naturally add a 'perhaps' to this. The connexion with F. will give my existence more strength to resist.

2. Everything immediately gives me pause. Every joke in the comic paper, what I remember about Flaubert and Grillparzer, the sight of the nightshirts on my parents' beds, laid out for the night, Max's marriage. Yesterday my sister said, 'All the married people (that we know) are happy, I don't understand it,' this remark too gave me pause, I became afraid again.

3. I must be alone a great deal. What I accomplished was only the result of being alone.

4. I hate everything that does not relate to literature, conversations bore me (even if they relate to literature), to visit people bores me, the sorrows and joys of my relatives bore me to my soul. Conversations take the importance, the seriousness, the truth of everything I think.

5. The fear of the connexion, of passing into the other. Then I'll never be alone again.

6. In the past, especially, the person I am in the company of my sisters has been entirely different from the person I am in the company of other people. Fearless, powerful, surprising, moved as I otherwise am only when I write. If through the intermediation of my wife I could be like that in the presence of everyone! But then would it not be at the expense of my writing? Not that, not that!

7. Alone, I could perhaps some day really give up my job. Married, it will never be possible.

FRANZ KAFKA, *The Diaries, 1910–23*

Kafka, though twice engaged to Felice Bauer in the period between 1912 and 1917, broke off the engagement each time and never married. He once wrote to her:

My health is only just good enough for myself alone, not good enough for marriage, let alone fatherhood.

I have a question for you alone, my brother: I throw this question like a plummet into your soul, to discover how deep it is.

You are young and desire marriage and children. But I ask you: are you a man who *ought* to desire a child?

Are you the victor, the self-conqueror, the ruler of your senses, the lord of your virtues? Thus I ask you.

Or do the animal and necessity speak from your desire? Or isolation? Or disharmony with yourself?

I would have your victory and your freedom long for a child. You should build living memorials to your victory and your liberation.

You should build beyond yourself. But first you must be built yourself, square-built in body and soul.

You should propagate yourself not only forward, but upward! May the garden of marriage help you to do it.

You should create a higher body, a first motion, a self-propelling wheel—you should create a creator.

Marriage: that I call the will of two to create the one who is more than those who created it. Reverence before one another, as before the willers of such a will—that I call marriage.

FRIEDRICH NIETZSCHE, *Thus Spoke Zarathustra*, 1883–92

I never was attached to that great sect,
Whose doctrine is, that each one should select
Out of the crowd a mistress or a friend,
And all the rest, though fair and wise, commend
To cold oblivion, though it is in the code
Of modern morals, and the beaten road
Which those poor slaves with weary footsteps tread,
Who travel to their home among the dead
By the broad highway of the world, and so
With one chained friend, perhaps a jealous foe,
The dreariest and the longest journey go.

PERCY BYSSHE SHELLEY, from 'Epipsychidion', 1821

Like a Dog with a bottle, fast ti'd to his tail,
Like Vermin in a trap, or a Thief in a Jail,
Or like a *Tory* in a Bog,
Or an Ape with a Clog:
Such is the man, who when he might go free,
Does his liberty loose,
For a Matrimony noose,
And sels himself into Captivity.
The Dog he do's howl, when his bottle do's jog,
The Vermin, the Theif, and the *Tory* in vain
Of the trap, of the Jail, of the Quagmire complain.
But welfare poor *Pug*! for he playes with his Clog;
And tho' he would be rid on't rather than his life,
Yet he lugg's it, and he hug's it, as a man does his wife.

The Second Part

How happy a thing were a wedding
And a bedding,
If a man might purchase a wife
For a twelve month, and a day;
But to live with her all a man's life,
For ever and for ay,
'Till she grow as grey as a Cat,
Good faith, Mr Parson, I thank you for that.

THOMAS FLATMAN, *The Batchelor's Song*, 1674

I told him, I had, perhaps, different notions of matrimony from what the received custom had given us of it; that I thought a woman was a free agent, as well as a man, and was born free, and could she manage herself suitably, might enjoy that liberty to as much purpose as the men do; that the laws of matrimony were indeed otherwise, and mankind at this time acted quite upon other principles; and those such that a woman gave herself entirely away from herself, in marriage, and capitulated only to be, at best, but an upper servant, and from the time she took the man, she was no better or worse than the servant among the Israelites, who had his ears bored, that is, nailed to the door-post, who by that act gave himself up to be a servant during life.

That the very nature of the marriage contract was, in short, nothing but giving up liberty, estate, authority, and everything to the man, and the woman was indeed a mere woman ever after, that is to say, a slave.

He replied, that though in some respects it was as I had said, yet I ought to consider that as an equivalent to this, the man had all the care of things devolved upon him; that the weight of business lay upon his shoulders, and as he had the trust, so he had the toil of life upon him; his was the labour, his the anxiety of living; that the woman had nothing to do but to eat the fat and drink the sweet; to sit still and look around her, be waited on and made much of, be served and loved, and made easy, especially if the husband acted as became him; and that, in general, the labour of the man was appointed to make the woman live quiet and unconcerned in the world; that they had the name of subjection without the thing; and if, in inferior families, they had the drudgery of the house, and care of the provisions upon them, yet they had, indeed, much the easier part; for in general, the women had only the care of managing, that is, spending what their husbands get; and that a woman had the name of subjection, indeed, but that they generally commanded, not the men only, but all they had; managed all for themselves; and where the man did his duty, the woman's life was all ease and tranquillity, and that she had nothing to do but to be easy, and to make all that were about her both easy and merry.

I returned, that while a woman was single, she was a masculine in her politic capacity; that she had then the full command of what she had, and the full direction of what she did; that she was a man in her separate capacity, to all intents and purposes that a man could be so to himself; that she was controlled by none, because accountable to none; and was in subjection to none.

*

I added, that whoever the woman was that had an estate, and would give it up to be the slave of a great man, that woman was a fool, and must be fit for nothing but a beggar; that it was my opinion a woman was as fit to govern and enjoy her own estate, without a man, as a man was without a woman; and that if she had a mind to gratify herself as to sexes, she might entertain a man as a man does a mistress; that while she was thus single she was her own, and if she gave away that power, she merited to be as miserable as it was possible that any creature could be.

All he could say could not answer the force of this as to argument, only this, that the other way was the ordinary method that the world was guided by; that he had reason to expect I should be content with that which all the world was contented with; that he was of the opinion, that a sincere affection between a man and his wife answered all the objections that I had made about the being a slave, a servant, and the like; and where there was a mutual love there could be no bondage, but that there was but one interest, one aim, one design, and all conspired to make both very happy.

DANIEL DEFOE, *Roxana*, 1724

If women are to effect a significant amelioration in their condition it seems obvious that they must refuse to marry.

GERMAINE GREER, *The Female Eunuch*, 1970

Many of us who consider ourselves card-carrying feminists also believe in the possibilities of marriage. There is more to marriage, we believe, than its function as a mechanism to preserve male power. Change the politics, adjust that balance of power, many feminists now argue, and marriage will no longer be sexist. In other words, don't bulldoze the institution—rehab it instead.

CARYL RIVERS, 1977

Furthermore, in choosing wives and husbands, they [the Utopians] observe earnestly and straightly a custom which seemed to us very fond and foolish. For a grave and honest matron sheweth the woman, be she maid or widow, naked to the wooer: and likewise a sage and discreet man exhibiteth the wooer naked to the woman. At this custom we laughed, and disallowed it as foolish. But they, on the other part, do greatly wonder at the folly of all other nations, which in *buying a colt* (whereas a little money is in hazard) be so chary and circumspect, that though he be almost all bare, yet they will not buy him, unless the saddle and all the harness be taken off—least under those coverings be hid some gall or sore. And yet in *choosing a wife*, which shall be either pleasure or displeasure to them all their life after, they be so rechless, that all the residue of the woman's body being covered with clothes, they esteem her scarcely by one hand breadth (for they can see no more but her face), and so to join her to them, not without great jeopardy of evil agreeing together—if any thing in her body afterward should chance to offend and mislike them.

For all men be not so wise as to have respect to the virtuous condition of the party. And the endowments of the body cause the virtues of the mind more to be esteemed and regarded: yea, even the marriages of wise men. Verily, so foul deformity may be hid under those coverings, that it may quite alienate and take away the man's mind from his wife, when it shall not be lawful for their bodies to be separate again. If such deformity happen by any chance after the marriage is consummate and finished, well: therein is no remedy but patience: every man must take his fortune well in worth. But it were well done that a law were made whereby all such deceits might be eschewed and avoided before hand.

THOMAS MORE, *Utopia*, 1516, trans. Raphe Robinson, 1551

I was ever of opinion that the honest man who married and brought up a large family did more service than he who continued single and only talked of population. From this motive I had scarcely taken orders a year, before I began to think seriously of matrimony, and chose my wife as she did her wedding-gown, not for a fine glossy surface, but such qualities as would wear well.

OLIVER GOLDSMITH, *The Vicar of Wakefield*, 1764

When he again talked of Mrs Careless . . . he said, 'If I had married her, it might have been as happy for me.' *Boswell.* 'Pray, Sir, do you not suppose that there are fifty women in the world, with any one of whom a man may be as happy, as with any one woman in particular?' *Johnson.* 'Ay, Sir, fifty thousand.' *Boswell.* 'Then, Sir, you are not of opinion with some who imagine that certain men and certain women are made for each other; and that they cannot be happy if they miss their counterparts.' *Johnson.* 'To be sure not, Sir. I believe marriages would in general be as happy, and often more so, if they were all made by the lord chancellor, upon a due consideration of the character and circumstances, without the parties having any choice in the matter.'

JAMES BOSWELL, *Life of Samuel Johnson*, 1791

'One [woman] is pretty nearly as good as another, as far as any judgment can be formed of them before marriage. It is only after marriage that they show their true qualities, as I know by bitter experience. Marriage is, therefore, a lottery, and the less choice and selection a man bestows on his ticket the better; for if he has incurred considerable pains and expense to obtain a lucky number, and his lucky number proves a blank he experiences not a simple, but a complicated disappointment; the loss of labour and money being superadded to the disappointment of drawing a blank, which, constituting simply and entirely the grievance of him who has chosen his ticket at random is, from its simplicity, the more endurable.'

THOMAS LOVE PEACOCK, *Nightmare Abbey*, 1818

People often say that marriage is an important thing, and should be much thought of in advance, and marrying people are cautioned that there are many who marry in haste and repent at leisure. I am not sure, however, that marriage may not be pondered over too much; nor do I feel certain that the leisurely repentance does not as often follow the leisurely marriages as it does the rapid ones. That some repent no one can doubt; but I am inclined to believe that most men and women take their lots as they find them, marrying as the birds do by force of nature, and going on with their mates with a general, though not perhaps an undisturbed satisfaction, feeling inwardly assured that Providence, if it have not done the very best for them, has done for them as well as they could do for themselves with all the thought in the world. I do not know that a woman can assure to herself, by her own prudence and taste, a good husband any more than she can add two cubits to her stature; but husbands have been made to be decently good,—and wives too, for the most part, in our country,—so that the thing does not require quite so much thinking as some people say.

ANTHONY TROLLOPE, *Can You Forgive Her?*, 1864

The young person of marriageable age does, of course, possess an ego-consciousness (girls more than men, as a rule), but, since he has only recently emerged from the mists of original unconsciousness, he is certain to have wide areas which still lie in the shadow and which preclude to that extent the formation of psychological relationship. This means, in practice, that the young man (or woman) can have only an incomplete understanding of himself and others, and is therefore imperfectly informed as to his, and their, motives. As a rule the motives he acts from are largely unconscious. Subjectively, of course, he thinks himself very conscious and knowing, for we constantly overestimate the existing content of consciousness, and it is a great and surprising discovery when we find that what we had supposed to be the final peak is nothing but the first step in a very long climb. The greater the area of unconsciousness, the less is marriage a matter of free choice, as is shown subjectively in the fatal compulsion one feels so acutely when one is in love.

C. G. JUNG, 'Marriage as a Psychological Relationship', 1925

Now and then she was so bewildered by discoveries that she came to wonder why she had married him, and why people do marry—really! The fact was that she had married him for the look in his eyes. It was a sad look, and beyond that it could not be described. Also, a little, she had married him for his bright untidy hair, and for that short oblique shake of the head which, with him, meant a greeting or an affirmative. She had not married him for his sentiments nor for his goodness of heart. Some points in him she did not like. He had a tendency to colds, and she hated him whenever he had a cold. She often detested his terrible tidiness, though it was a convenient failing.... She did not like his way of walking, which was ungainly, nor his way of standing, which was infirm. She preferred him to be seated. She could not but regret his irresolution and his love of ease. However, the look in his eyes was paramount, because she was in love with him. She knew that he was more deeply and helplessly in love with her than she was with him, but even she was perhaps tightlier bound than in her pride she thought.

ARNOLD BENNETT, *These Twain*, 1916

I have come to be very much of a cynic in these matters; I mean that it is impossible to believe in the permanence of man's or woman's love. Or, at any rate, it is impossible to believe in the permanence of any early passion. As I see it, at least, with regard to man, a love affair, a love for any definite woman, is something in the nature of a widening of the experience. With each new woman that a man is attracted to there appears to come a broadening of the outlook, or, if you like, an acquiring of new territory. A turn of the eyebrow, a tone of the voice, a queer characteristic gesture—all these things, and it is these things that cause to arise the passion of love—all these things are like so many objects on the horizon of the landscape that tempt a man to walk beyond the horizon, to explore. He wants to get, as it were, behind those eyebrows with the peculiar turn, as if he desired to see the world with the eyes that they overshadow. He wants to hear that voice applying itself to every possible proposition, to every possible topic; he wants to see those characteristic gestures against every possible background. Of the question of the sex instinct I know very little and I do not think that it counts for very much in a really great passion. It can be aroused by such nothings—by an untied shoelace, by a glance of the eye in passing—that I think it might be left out of the calculation. I don't mean to say that any great passion can exist without a desire for consummation. That seems to me to be a commonplace and to be therefore a matter needing no comment at all. It is a thing, with all its accidents, that must be taken for granted, as in a novel, or a biography, you take it for granted that the characters have their meals with some regularity. But the real fierceness of desire, the real heat of a passion long continued and withering up the soul of a man, is the craving for identity with the woman that he loves. He desires to see with the same eyes, to touch with the same sense of touch, to hear with the same ears, to lose his identity, to be enveloped, to be supported. For, whatever may be said of the relation of the sexes, there is no man who loves a woman that does not desire to come to her for the renewal of his courage, for the cutting asunder of his difficulties. And that will be the mainspring of his desire for her. We are all so afraid, we are all so alone, we all so need from the outside the assurance of our own worthiness to exist.

And yet I do believe that for every man there comes at last a woman—or, no, that is the wrong way of formulating it. For every man there comes at last a time of life when the woman who then sets her seal upon his imagination has set her seal for good. He will travel over no more horizons; he will never again set the knapsack over his shoulders; he will retire from those scenes. He will have gone out of the business.

FORD MADOX FORD, *The Good Soldier*, 1915

MARRIAGE I THINK

Marriage I think
For women
Is the best of opiates.
It kills the thoughts
That think about the thoughts,
It is the best of opiates.
So said Maria.
But too long in solitude she'd dwelt,
And too long her thoughts had felt
Their strength. So when the man drew near,
Out popped her thoughts and covered him with fear.
Poor Maria!
Better that she had kept her thoughts on a chain,
For now she's alone again and all in pain;
She sighs for the man that went and the thoughts that stay
To trouble her dreams by night and her dreams by day.

STEVIE SMITH, 1937

'Ursula,' said Gudrun, 'don't you *really want* to get married?' Ursula laid her embroidery in her lap and looked up. Her face was calm and considerate.

'I don't know,' she replied. 'It depends how you mean.'

Gudrun was slightly taken aback. She watched her sister for some moments.

'Well,' she said, ironically, 'it usually means one thing! But don't you think, anyhow, you'd be—' she darkened slightly—'in a better position than you are in now.'

A shadow came over Ursula's face.

'I might,' she said. 'But I'm not sure.'

Again Gudrun paused, slightly irritated. She wanted to be quite definite.

'You don't think one needs the *experience* of having been married?' she asked.

'Do you think it need *be* an experience?' replied Ursula.

'Bound to be, in some way or other,' said Gudrun, coolly. 'Possibly undesirable, but bound to be an experience of some sort.'

'Not really,' said Ursula. 'More likely to be the end of experience.'

Gudrun sat very still, to attend to this.

'Of course,' she said, 'there's *that* to consider.' This brought the conversation to a close. Gudrun, almost angrily, took up her rubber and began to rub out part of her drawing. Ursula stitched absorbedly.

'You wouldn't consider a good offer?' asked Gudrun.

'I think I've rejected several,' said Ursula.

'*Really!*' Gudrun flushed dark—'But anything really worth while? Have you *really*?'

'A thousand a year, and an awfully nice man. I liked him awfully,' said Ursula.

'Really! But weren't you fearfully tempted?'

'In the abstract but not in the concrete,' said Ursula. 'When it comes to the point, one isn't even tempted—oh, if I were tempted, I'd marry like a shot. I'm only tempted *not* to.' The faces of both sisters suddenly lit up with amusement.

'Isn't it an amazing thing,' cried Gudrun, 'how strong the temptation is, not to!' They both laughed, looking at each other. In their hearts they were frightened.

There was a long pause, whilst Ursula stitched and Gudrun went on with her sketch. The sisters were women, Ursula twenty-six, and Gudrun twenty-five. But both had the remote, virgin look of modern girls, sisters of Artemis rather than of Hebe.

*

'Don't you find yourself getting bored?' she asked of her sister. 'Don't you find, that things fail to materialize? *Nothing materializes!* Everything withers in the bud.'

'What withers in the bud?' asked Ursula.

'Oh, everything—oneself—things in general.' There was a pause, whilst each sister vaguely considered her fate.

'It does frighten one,' said Ursula, and again there was a pause. 'But do you hope to get anywhere by just marrying?'

'It seems to be the inevitable next step,' said Gudrun. Ursula pondered this, with a little bitterness.

*

'I know,' she said, 'it seems like that when one thinks in the abstract. But really imagine it: imagine any man one knows, imagine him coming home to one every evening, and saying "Hello", and giving one a kiss—'

There was a blank pause.

'Yes,' said Gudrun, in a narrowed voice. 'It's just impossible. The man makes it impossible.'

D. H. LAWRENCE, *Women in Love*, 1921

You'll say the gentleman is somewhat simple—
The better for a husband, were you wise,
For those that marry fools live ladies' lives.

THOMAS MIDDLETON, *Women Beware Women*, 1657

'He is a chump, you know. That's what I love about him. That and the way his ears wiggle when he gets excited. Chumps always make the best husbands. When you marry, Sally, grab a chump. Tap his forehead first, and if it rings solid, don't hesitate. All the unhappy marriages come from the husband having brains. What good are brains to a man? They only unsettle him.'

P. G. WODEHOUSE, *The Adventures of Sally*, 1922

But who is to tell whether a girl will make an industrious woman? How is the purblind lover especially, to be able to ascertain whether she, whose smiles and dimples, and bewitching lips have half bereft him of his senses; how is he able to judge, from anything that he can see, whether the beloved object will be industrious or lazy? Look a little at the labours of the teeth, for these correspond with those of the other members of the body, and with the operations of the mind. 'Quick at meals, quick at work,' is a saying as old as the hills, in this, the most industrious nation upon earth; and never was there a truer saying. But fashion comes in here, and decides that you shall not be quick at meals; that you shall sit and be carrying on the affair of eating for an hour or more. Good God! what have I not suffered on this account! However, though she must sit as long as the rest and though she join in the performance (for it is a real performance) unto the end of the last scene, she cannot make her teeth abandon their character. She may, and must suffer the slice to linger on the plate, and must make the supply slow, in order to fill up the time; but when she does bite, she cannot well disguise what nature has taught her to do; and you may be assured, that if her jaws move in slow time, and if she rather squeeze than bite the food; if she so deal with it as to leave you in doubt as to whether she mean finally to admit or reject it; if she deal with it thus, set her down as being, in her nature, incorrigibly lazy. Never mind the pieces of needlework, the tambouring, the maps of the world made by her needle. Get to see her at work upon a mutton-chop, or a bit of bread and cheese; and if she deal quickly with these, you have a pretty good security for that activity, that stirrng industry, without which a wife is a burden instead of being a help. And, as to love, it cannot live for more than a month or two (in the breast of a man of spirit) towards a lazy woman.

WILLIAM COBBETT, *Advice to Young Men*, 1830

The determinants of women's choice of an object are often made unrecognizable by social conditions. Where the choice is able to show itself freely, it is often made in accordance with the narcissistic ideal of the man whom the girl had wished to become.

SIGMUND FREUD, *New Introductory Lectures on Psychoanalysis*, 1933

ANN: Violet is quite right. You ought to get married.

TANNER [*explosively*]: Ann: I will not marry you. Do you hear? I won't, won't, won't, won't, WON'T marry you.

ANN [*placidly*]: Well, nobody axd you, sir she said, sir she said, sir she said. So that's settled.

TANNER: Yes, nobody has asked me; but everybody treats the thing as settled. It's in the air.

*

ANN: Well, if you don't want to be married, you needn't be [*she turns away from him and sits down, much at her ease*].

TANNER [*following her*]: Does any man want to be hanged? Yet men let themselves be hanged without a struggle for life, though they could at least give the chaplain a black eye. We do the world's will, not our own. I have a frightful feeling that I shall let myself be married because it is the world's will that you should have a husband.

ANN: I daresay I shall, someday.

TANNER: But why me? me of all men! Marriage is to me apostasy, profanation of the sanctuary of my soul, violation of my manhood, sale of my birthright, shameful surrender, ignominious capitulation, acceptance of defeat. I shall decay like a thing that has served its purpose and is done with; I shall change from a man with a future to a man with a past; I shall see in the greasy eyes of all the other husbands their relief at the arrival of a new prisoner to share their ignominy. The young men will scorn me as one who has sold out: to the women I, who have always been an enigma and a possibility, shall be merely somebody else's property—and damaged goods at that: a secondhand man at best.

ANN: Well, your wife can put on a cap and make herself ugly to keep you in countenance, like my grandmother.

TANNER: So that she may make her triumph more insolent by publicly throwing away the bait the moment the trap snaps on the victim!

ANN: After all, though, what difference would it make? Beauty is all very well at first sight; but who ever looks at it when it has been in the house three days? I thought our pictures very lovely when papa bought them; but I haven't looked at them for years. You never bother about my looks: you are too well used to me. I might be the umbrella stand.

TANNER: You lie, you vampire: you lie.

ANN: Flatterer. Why are you trying to fascinate me, Jack, if you don't want to marry me?

TANNER: The Life Force. I am in the grip of the Life Force.

ANN: I don't understand in the least: it sounds like the Life Guards.

TANNER: Why don't you marry Tavy? He is willing. Can you not be satisfied unless your prey struggles?

ANN [*turning to him as if to let him into a secret*]: Tavy will never marry. Haven't you noticed that that sort of man never marries? . . . men like that always live in comfortable bachelor lodgings with broken hearts, and are adored by their landladies, and never get married. Men like you always get married.

TANNER [*smiting his brow*]: How frightfully, horribly true! It has been staring me in the face all my life; and I never saw it before.

ANN: Oh, it's the same with women. The poetic temperament's a very nice temperament, very amiable, very harmless and poetic, I daresay; but it's an old maid's temperament.

TANNER: Barren. The Life Force passes it by.

ANN: If that's what you mean by the Life Force, yes.

TANNER: You don't care for Tavy?

ANN [*looking round carefully to make sure that Tavy is not within earshot*]: No.

TANNER: And you do care for me?

ANN [*rising quietly and shaking her finger at him*]: Now, Jack! Behave yourself.

TANNER: Infamous, abandoned woman! Devil!

ANN: Boa-constrictor! Elephant!

TANNER: Hypocrite!

ANN [*softly*]: I must be, for my future husband's sake.

TANNER: For mine! [*Correcting himself savagely*] I mean for his.

ANN [*ignoring the correction*]: Yes, for yours. You had better marry what you call a hypocrite, Jack. Women who are not hypocrites go about in rational dress and are insulted and get into all sorts of hot water. And then their husbands get dragged in too, and live in continual dread of fresh complications. Wouldn't you prefer a wife you could depend on?

TANNER: No: a thousand times no: hot water is the revolutionist's element. You clean men as you clean milkpails, by scalding them.

ANN: Cold water has its uses too. It's healthy.

TANNER [*despairingly*]: Oh, you are witty: at the supreme moment the Life Force endows you with every quality. Well, I too can be a hypocrite. Your father's will appointed me your guardian, not your suitor. I shall be faithful to my trust.

ANN [*in low siren tones*]: He asked me who I would have as my guardian before he made that will. I chose you!

TANNER: The will is yours then! The trap was laid from the beginning.

ANN [*concentrating all her magic*]: From the beginning—from our childhood—for both of us—by the Life Force.

GEORGE BERNARD SHAW, *Man and Superman*, 1903

The beginning—to which she often went back—had been a scene, for our young woman, of supreme brilliancy; a party given at a 'gallery' hired by a hostess who fished with big nets.

*

They had found themselves regarding each other straight, and for a longer time on end than was usual even at parties in galleries; but that in itself after all would have been a small affair for two such handsome persons. It wasn't, in a word, simply that their eyes had met; other conscious organs, faculties, feelers had met as well, and when Kate afterwards imagined to herself the sharp deep fact she saw it, in the oddest way, as a particular performance. She had observed a ladder against a garden-wall and had trusted herself so to climb it as to be able to see over into the probable garden on the other side. On reaching the top she had found herself face to face with a gentleman engaged in a like calculation at the same moment, and the two enquirers had remained confronted on their ladders. The great point was that for the rest of that evening they had been perched—they had not climbed down; and indeed during the time that followed Kate at least had a perched feeling—it was as if she were there aloft without a retreat.

HENRY JAMES, *The Wings of a Dove*, 1902

Only a few days ago—and yet the day seems already to belong to a remote past, and to be separated from these last dark hours by a great gulf, misty, not to be passed,—I realised that a new power had come into my life—the heavenly power that makes all things new. I had gone down to the cottage in a hot, breathless sunlit afternoon. I had long passed the formality of ringing to announce my entrance. There was no one in the little drawing-room, which was cool and dark, with shuttered windows. I went out upon the lawn. Miss Waring was sitting in a chair under a beech tree reading, and at the sight of me she rose, laid down her book, and came smiling across the grass. There is a subtle, viewless message of the spirit which flashes between kindred souls, in front of and beyond the power of look or speech, and at the same moment that I understood I felt she understood too. I could not then at once put into words my hopes; but it hardly seemed necessary. We sat together, we spoke a little, but were mostly silent in some secret interchange of spirit. That afternoon my heart climbed, as it were, a great height, and saw from a Pisgah top the familiar land at its feet, all lit with a holy radiance, and then turning, saw, in golden gleams and purple haze, the margins of an unknown sea stretching out beyond the sunset to the very limits of the world.

H. C. BENSON, *The House of Quiet*, 1904

THE CONFIRMATION

Yes, yours, my love, is the right human face.
I in my mind had waited for this long,
Seeing the false and searching for the true,
Then found you as a traveller finds a place
Of welcome suddenly amid the wrong
Valleys and rocks and twisting roads. But you,
What shall I call you? A fountain in a waste,
A well of water in a country dry,
Or anything that's honest and good, an eye
That makes the whole world bright. Your open heart,
Simple with giving, gives the primal deed,
The first good world, the blossom, the blowing seed,
The hearth, the steadfast land, the wandering sea,
Not beautiful or rare in every part,
But like yourself, as they were meant to be.

EDWIN MUIR, 1943

Campaigners against [marriage], from Shelley and the Mills on, have been remarkably crass in posing the simple dilemma, 'either you want to stay together or you don't—if you do, you need not promise; if you don't, you ought to part.' This ignores the chances of inner conflict, and the deep human need for a continuous central life that lasts through genuine, but passing, changes of mood. The need to be able to rely on other people is not some sort of shameful weakness; it is an aspect of the need to be true to oneself.

MARY MIDGLEY, *Beast and Man*, 1978

THE GOOD MORROW

I wonder by my troth, what thou and I
Did, till we loved? were we not weaned till then?
But sucked on country pleasures, childishly?
Or snorted we i'the seven sleepers' den?
'Twas so: But this, all pleasures fancies be.
If ever any beauty I did see,
Which I desired, and got, 'twas but a dream of thee.
And now good morrow to our waking souls,
Which watch not one another out of fear;
For love, all love of other sights controls,
And makes one little room, an everywhere.
Let sea-discoverers to new worlds have gone,
Let maps to others, worlds on worlds have shown,
Let us possess our world, each hath one, and is one.

My face in thine eye, thine in mine appears,
And true plain hearts do in the faces rest,
Where can we find two better hemispheres
Without sharp North, without declining West?
Whatever dies, was not mixed equally;
If our two loves be one, or, thou and I
Love so alike, that none do slacken, none can die.

JOHN DONNE, 1633

2

Marry me, Marry me, Marry me

PROPOSALS—breathlessly passionate and breathtakingly indifferent, ardently idealistic and sternly practical, tremulously romantic and insolently patronizing, poignantly inarticulate and urbanely stylish, painfully serious and elegantly witty: what a pity that a convention which allows of such infinite variety seems barely to have survived the change in sexual mores, or the advent of the telephone.

Tradition has it that men do the asking—sometimes at great cost, as Charlotte Brontë recognized—but even in the eighteenth and nineteenth centuries, despite the stylized manners of the time, it is often the woman who initiates or invites the proposal. Only the inappropriate proposal is truly 'so sudden'. Mr Guppy's and Mr Collins's unrealistic views of themselves as potential marriage partners is light-years removed from the awareness that E. M. Forster's heroine has of 'the central radiance', or the sympathetic understanding between Kitty and Levin, who have no need of explicit proposal or acceptance.

Perhaps the best proposals, like the best decisions, never need to be made, but become self-evident. Not only is the quintessential nature of each partner revealed in this highly charged moment, but one could probably predict the course of a marriage from the way in which the couple take this crucial step.

One partner's conviction, based on intuition or conscious self-knowledge, may suffice to make the marriage seem inevitable. The need for a 'high-hatted' lover of the opening verse, or Willoughby's insistence that Laetitia, who is his severest critic, *must* marry him, are examples of such unilateral recognition.

In conventional romantic fiction the proposal comes at the end of the story: in real life it is only the beginning. As Ved Mehta's father says, love is not an act, like falling in love; it is a lifelong process.

Then wear the gold hat, if that will move her;
  If you can bounce high, bounce for her too,
Till she cry 'Lover, gold-hatted, high-bouncing lover,
  I must have you!'

THOMAS PARKE D'INVILLIERS

¶ *This is the epigraph to Scott Fitzgerald's* The Great Gatsby *(1925). There is no record of d'Invilliers. Perhaps Fitzgerald invented him?*

MR EDWARDS: Myfanwy Price!
MISS PRICE: Mr Mog Edwards!
MR EDWARDS: I am a draper mad with love. I love you more than all the flannelette and calico, candlewick, dimity, crash and merino, tussore, cretonne, crepon, muslin, poplin, ticking and twill in the whole Cloth Hall of the world. I have come to take you away to my Emporium on the hill, where the change hums on wires. Throw away your little bedsocks and your Welsh wool knitted jacket, I will warm the sheets like an electric toaster, I will lie by your side like the Sunday roast.
MISS PRICE: I will knit you a wallet of forget-me-not blue, for the money to be comfy. I will warm your heart by the fire so that you can slip it in under your vest when the shop is closed.
MR EDWARDS: Myfanwy, Myfanwy, before the mice gnaw at your bottom drawer will you say
MISS PRICE: Yes, Mog, yes, Mog, yes, yes, yes.
MR EDWARDS: And all the bells of the tills of the town shall ring for our wedding.

DYLAN THOMAS, *Under Milk Wood*, 1954

Quoth John to Joan, will thou have me:
I prithee now, wilt? and I'll marry thee,
My cow, my calf, my house, my rents,
And all my lands and tenements:

Oh, say, my Joan, will not that do?
I cannot come every day to woo.

I've corn and hay in the barn hard-by,
And three fat hogs pent up in the sty,
I have a mare and she is coal black,
I ride on her tail to save my back.
Then, say, my Joan, will not that do?
I cannot come every day to woo.

I have a cheese upon the shelf,
And I cannot eat it all myself;
I've three good marks that lie in a rag,
In a nook of the chimney, instead of a bag.
Then, say, my Joan, will not that do?
I cannot come every day to woo.

To marry I would have thy consent,
But faith I never could compliment;
I can say nought but 'Hoy, gee ho!'
Words that belong to the cart and the plough.
Oh, say, my Joan, will not that do?
I cannot come every day to woo.

ANON.

A love letter, written in 1712:

Lovely (and oh, that I could write loving) Mistress Margaret Clark, I pray you let affection excuse presumption. Having been so happy as to enjoy the sight of your sweet countenance, I am so enamoured with you that I can no more keep close my flaming desire to become your servant.

And I am the more bold now to write to your sweet self because I am now my own man, and may match where I please, for my father is taken away, and now I am come to my living, which is Ten Yard Land and a House, and there is never a yard of land in our field but is well worth ten pounds a year; and all my brothers and sisters are provided for.

Besides I have good household stuff, though I say it, both of brass and pewter, linens and woollens; and, though my house be thatched, it shall go hard but I will have one half of it slated.

If you think well of this motion I will wait upon you as soon as my new clothes are made and the hay harvest is in.

from ARTHUR MEE: *One Thousand Beautiful Things*, 1925

Sukey, you shall be my wife
And I will tell you why:
I have got a little pig,
And you have got a sty;
I have got a dun cow
And you can make good cheese;
Sukey, will you marry me?
Say Yes, if you please.

ANON., nursery rhyme

HOPEFUL PROPOSAL TO A YOUNG LADY OF THE VILLAGE,

Dated NOVEMBER 29th, 1866

My Dear Miss,
I now take up my pen to write to you hoping these few lines will find you well as it leaves me at present Thank God for it. You will perhaps be surprised that I should make so bold as to write to you who is such a lady and I hope you will not be vex at me for it. I hardly dare say what I want, I am so timid about ladies, and my heart trimmels like a hespin. But I once seed in a book that faint heart never won fair lady, so here goes.

I am a farmer in a small way and my age is rather more than forty years and my mother lives with me and keeps my house, and she has been very poorly lately and cannot stir about much and I think I should be more comfortabler with a wife.

I have had my eye on you a long time and I think you are a very nice young woman and one that would make me happy if only you think so. We keep a servant girl to milk three kye and do the work in the house, and she goes out a bit in the summer to gadder wickens and she snags a few of turnips in the back kend. I do a piece of work on the farm myself and attends Pately Market, and I sometimes show a few sheep and I feeds between 3 & 4 pigs agen Christmas, and the same is very useful in the house to make pies and cakes and so forth, and I sells the hams to help pay for the barley meal.

I have about 73 pund in Naisbro Bank and we have a nice little parlour downstairs with a blue carpet, and an oven on the side of the fireplace and the old woman on the other side smoking. The Golden Rules claimed up on the walls above the long settle, and you could sit all day in the easy chair and knit and mend my kytles and leggums, and you could make the tea ready agin I come in, and you could make butter for Pately Market, and I would drive you to church every Sunday in the spring cart, and I would do all that bees in my power to make you happy. So I hope to hear from you. I am in desprit and Yurnest, and will marry you at May Day, or if my mother dies afore I shall want you afore. If only you will accept of me, my dear, we could be very happy together.

I hope you will let me know your mind by return of post, and if you are favourable I will come up to scratch. So no more at present from your well-wisher and true love—

Simon Fallowfield

P.S. I hope you will say nothing about this. If you will not accept of me I have another very nice woman in my eye, and I think I shall marry her if you do not accept of me, but I thought you would suit me mother better, she being very crusty at times. So I tell you now before you come, she will be Maister.

from Steve Race's family archive

¶ *A proposal received (and refused!) by Mary Foster, the local beauty of Middlemoor, Pateley Bridge, in Yorkshire.*

He turned upon Meriam, who shrank back in terror.

'Come here—you. I'll take you instead. Ay, dirt as you are, I'll take you, and we'll sink into th' mud together. There have always been Starkadders at Cold Comfort, and now there'll be a Beetle too.'

'And not the first neither, as you'd know if you'd ever cleaned out the larder,' said a voice, tartly. It was Mrs Beetle herself, who, hitherto unobserved by Flora, had been busily cutting bread and butter and replenishing the glasses of the farm-hands in a far corner of the long kitchen. She now came forward into the circle about the fire, and confronted Urk with her arms akimbo.

'Well . . . 'oo's talking about dirt? 'Eaven knows, you should know something about it, in that coat and them trousers. Enough ter turn up one of yer precious water-voles, you are. A pity you don't spend a bit less time with yer old water-voles and a bit more with a soap and flannel.'

Here she received unexpected support from Mark Dolour, who called in a feeling tone from the far end of the kitchen:

'Ay, that's right.'

'Don't you 'ave 'im, ducky, unless you feel like it,' advised Mrs Beetle, turning to Meriam. 'You're full young yet, and 'e won't see forty again.'

'I don't mind. I'll 'ave him, if 'un wants me,' said Meriam, amiably. 'I can always make 'im wash a bit, if I feels like it.'

Urk gave a wild laugh. His hand fell on her shoulder, and he drew her to him and pressed a savage kiss full on her open mouth. Aunt Ada Doom, choking with rage, struck at them with the 'Milk Producers' Weekly Bulletin and Cowkeepers' Guide,' but the blow missed. She fell back, gasping, exhausted.

'Come, my beauty—my handful of dirt. I mun carry thee up to Ticklepenny's and show 'ee to the water-voles.' Urk's face was working with passion.

'What! At this time o' night?' cried Mrs Beetle, scandalised.

Urk put one arm round Meriam's waist and heaved away, but could not budge her from the floor. He cursed aloud, and, kneeling down, placed his arms about her middle, and heaved again. She did not stir. Next he wrapped his arms about her shoulders, and below her knees. She declined upon him, and he, staggering beneath her, sank to the floor. Mrs Beetle made a sound resembling 't-t-t-t-t.'

Mark Dolour was heard to mutter that th' Fireman's Lift was as good a hold as any he knew.

Now Urk made Meriam stand in the middle of the floor, and with a low, passionful cry, ran at her.

'Come, my beauty.'

The sheer animal weight of the man bore her up into his clutching arms. Mark Dolour (who dearly loved a bit of sport) held open the door, and Urk and his burden rushed out into the dark and the earthy scents of the young spring night.

STELLA GIBBONS, *Cold Comfort Farm*, 1933

My dear Sir. This letter comes to know whether you will be pleased to give me leave to propose marriage to your daughter, Miss Elizabeth. You need not be afraid of sending me a refusal; for I bless God, if I know anything of my own heart, I am free from that foolish passion which the world calls love.

¶ *from a letter by George Whitefield (1714–70), founder of Calvinistic Methodism in England, and a powerful preacher. He married Elisabeth James, a widow ten years his senior, who 'had neither fortune nor beauty'.*

AUTUMN

He told his life story to Mrs Courtly
Who was a widow. 'Let us get married shortly',
He said. 'I am no longer passionate,
But we can have some conversation before it is too late.'

STEVIE SMITH, (1942)

THE COURTSHIP OF THE YONGHY-BONGHY-BÒ

On the Coast of Coromandel
  Where the early pumpkins blow,
    In the middle of the woods
  Lived the Yonghy-Bonghy-Bò.
Two old chairs, and half a candle,—
One old jug without a handle,—
    These were all his worldly goods:
    In the middle of the woods,
    These were all the worldly goods,
  Of the Yonghy-Bonghy-Bò,
  Of the Yonghy-Bonghy-Bò.

Once, among the Bong-trees walking
  Where the early pumpkins blow,
    To a little heap of stones
  Came the Yonghy-Bonghy-Bò.
There he heard a Lady talking,
To some milk-white Hens of Dorking,—

''Tis the Lady Jingly Jones!
'On that little heap of stones
'Sits the Lady Jingly Jones!'
Said the Yonghy-Bonghy-Bò.
Said the Yonghy-Bonghy-Bò.

'Lady Jingly! Lady Jingly!
'Sitting where the pumpkins blow,
'Will you come and be my wife?'
Said the Yonghy-Bonghy-Bò.
'I am tired of living singly,–
'On this coast so wild and shingly,—
'I'm a-weary of my life;
'If you'll come and be my wife,
'Quite serene would be my life!'—
Said the Yonghy-Bonghy-Bò,
Said the Yonghy-Bonghy-Bò.

'On this Coast of Coromandel,
'Shrimps and watercresses grow,
'Prawns are plentiful and cheap,'
Said the Yonghy-Bonghy-Bò.
'You shall have my chairs and candle,
'And my jug without a handle!—
'Gaze upon the rolling deep
'(Fish is plentiful and cheap);
'As the sea, my love is deep!'
Said the Yonghy-Bonghy-Bò.
Said the Yonghy-Bonghy-Bò.

Lady Jingly answered sadly,
And her tears began to flow,—
'Your proposal comes too late,
'Mr Yonghy-Bonghy-Bò!
'I would be your wife most gladly!'
(Here she twirled her fingers madly)
'But in England I've a mate!
'Yes! you've asked me far too late,
'For in England I've a mate,
'Mr Yonghy-Bonghy-Bò!
'Mr Yonghy-Bonghy-Bò!

*

'Though you've such a tiny body,
'And your head so large doth grow,—
'Though your hat may blow away,
'Mr Yonghy-Bonghy-Bò!
'Though you're such a Hoddy Doddy—
'Yet I wish that I could modi-
'fy the words I needs must say!
'Will you please to go away?
'That is all I have to say—
'Mr Yonghy-Bonghy-Bò!
'Mr Yonghy-Bonghy-Bò!'

*

From the Coast of Coromandel,
Did that Lady never go;
On that heap of stones she mourns
For the Yonghy-Bonghy-Bò.
On that Coast of Coromandel,
In his jug without a handle,
Still she weeps, and daily moans;
On that little heap of stones
To her Dorking Hens she moans,
For the Yonghy-Bonghy-Bò,
For the Yonghy-Bonghy-Bò.

EDWARD LEAR, 1877

On finding Mrs Bennet, Elizabeth, and one of the younger girls together, soon after breakfast, [Mr Collins] addressed the mother in these words,—

'May I hope, madam, for your interest with your fair daughter Elizabeth, when I solicit for the honour of a private audience with her in the course of this morning?'

Before Elizabeth had time for anything but a blush of surprise, Mrs Bennet instantly answered,—

'Oh dear! Yes, certainly. I am sure Lizzy will be very happy; I am sure she can have no objection.—Come, Kitty; I want you upstairs.' And gathering her work together, she was hastening away, when Elizabeth called out,—

'Dear ma'am, do not go. I beg you will not go. Mr Collins must excuse me. He can have nothing to say to me that anybody need not hear. I am going away myself.'

'No, no; nonsense, Lizzy. I desire you will stay where you are.' And upon Elizabeth's seeming really, with vexed and embarrassed looks, about to escape, she added, 'Lizzy, I *insist* upon your staying and hearing Mr Collins.'

Elizabeth would not oppose such an injunction; and a moment's consideration making her also sensible that it would be wisest to get it over as soon and as quietly as possible, she sat down again, and tried to conceal, by incessant employment, the feelings which were divided between distress and diversion. Mrs Bennet and Kitty walked off, and as soon as they were gone Mr Collins began,—

'Believe me, my dear Miss Elizabeth, that your modesty, so far from doing you any disservice, rather adds to your other perfections. You would have been less amiable in my eyes had there *not* been this little unwillingness; but allow me to assure you that I have your respected mother's permission for this address. You can hardly doubt the purport of my discourse, however your natural delicacy may lead you to dissemble; my attentions have been too marked to be mistaken. Almost as soon as I entered the house I singled you out as the companion of my future life. But before I am run away with by my feelings on this subject, perhaps it will be advisable for me to state my reasons for marrying, and, moreover, for coming into Hertfordshire with the design of selecting a wife, as I certainly did.'

The idea of Mr Collins, with all his solemn composure, being run away with by his feelings, made Elizabeth so near laughing that she could not use the short pause he allowed in any attempt to stop him further, and he continued,—

'My reasons for marrying are, first, that I think it a right thing for every clergyman in easy circumstances (like myself) to set the example of matrimony in his parish; secondly, that I am convinced it will add very greatly to my happiness; and thirdly, which perhaps I ought to have mentioned earlier, that it is the particular advice and recommendation of the very noble lady whom I have the honour of calling patroness. Twice has she condescended to give me her opinion (unasked too!) on this subject; and it was but the very Saturday night before I left Hunsford—between our pools at quadrille, while Mrs Jenkinson was arranging Miss De Bourgh's footstool—that she said, "Mr Collins, you must marry. A clergyman like you must marry.

Choose properly, choose a gentlewoman, for *my* sake and for your *own*; let her be an active, useful sort of person, not brought up high, but able to make a small income go a good way. This is my advice. Find such a woman as soon as you can, bring her to Hunsford, and I will visit her." Allow me, by the way, to observe, my fair cousin, that I do not reckon the notice and kindness of Lady Catherine de Bourgh as among the least of the advantages in my power to offer. You will find her manners beyond anything I can describe; and your wit and vivacity, I think, must be acceptable to her, especially when tempered with the silence and respect which her rank will inevitably excite. Thus much for my general intention in favour of matrimony; it remains to be told why my views were directed to Longbourn instead of my own neighbourhood, where, I assure you, there are many amiable young women. But the fact is, that being, as I am, to inherit this estate after the death of your honoured father (who, however, may live many years longer), I could not satisfy myself without resolving to choose a wife from among his daughters, that the loss to them might be as little as possible when the melancholy event takes place which, however, as I have already said, may not be for several years. This has been my motive, my fair cousin, and I flatter myself it will not sink me in your esteem. And now nothing remains for me but to assure you in the most animated language of the violence of my affection. To fortune I am perfectly indifferent, and shall make no demand of that nature on your father, since I am well aware that it could not be complied with, and that one thousand pounds in the four per cents., which will not be yours till after your mother's decease, is all that you may ever be entitled to. On that head, therefore, I shall be uniformly silent, and you may assure yourself that no ungenerous reproach shall ever pass my lips when we are married.'

It was absolutely necessary to interrupt him now.

'You are too hasty, sir,' she cried. 'You forget that I have made no answer. Let me do it without further loss of time. Accept my thanks for the compliment you are paying me. I am very sensible of the honour of your proposals, but it is impossible for me to do otherwise than decline them.'

'I am not now to learn,' replied Mr Collins, with a formal wave of the hand, 'that it is usual with young ladies to reject the addresses of the man whom they secretly mean to accept, when he first applies for their favour; and that sometimes the refusal is repeated a second or even a third time. I am, therefore, by no means discouraged by what you have just said, and shall hope to lead you to the altar ere long.'

'Upon my word, sir,' cried Elizabeth, 'your hope is rather an extraordinary one after my declaration. I do assure you that I am not one of those young ladies (if such young ladies there are) who are so daring as to risk their happiness on the chance of being asked a second time. I am perfectly serious in my refusal. You could not make *me* happy, and I am convinced that I am the last woman in the world who would make *you* so. Nay, were your friend Lady Catherine to know me, I am persuaded she would find me in every respect ill qualified for the situation.'

'Were it certain that Lady Catherine would think so,' said Mr Collins, very gravely; 'but I cannot imagine that her ladyship would at all disapprove of you. And you may be certain that when I have the honour of seeing her again I shall speak in the highest terms of your modesty, economy, and other amiable qualifications.'

'Indeed, Mr Collins, all praise of me will be unnecessary. You must give me leave to judge for myself, and pay me the compliment of believing what I say. I wish you very happy and very rich, and by refusing your hand do all in my power to prevent your being otherwise. In making me the offer, you must have satisfied the delicacy of your feelings with regard to my family, and may take possession of Longbourn estate whenever it falls, without any self-reproach. This matter may be considered, therefore, as finally settled.' And rising as she thus spoke, she would have quitted the room, had not Mr Collins thus addressed her,—

'When I do myself the honour of speaking to you next on the subject, I shall hope to receive a more favourable answer than you have now given me; though I am far from accusing you of cruelty at present, because I know it to be the established custom of your sex to reject a man on the first application, and perhaps you have even now said as much to encourage my suit as would be consistent with the true delicacy of the female character.'

'Really, Mr Collins,' cried Elizabeth, with some warmth, 'you puzzle me exceedingly. If what I have hitherto said can appear to you in the form of encouragement, I know not how to express my refusal in such a way as may convince you of its being one.'

'You must give me leave to flatter myself, my dear cousin, that your refusal of my addresses are merely words of course. My reasons for believing it are briefly these:—It does not appear to me that my hand is unworthy your acceptance, or that the establishment I can offer would be any other than highly desirable. My situation in life, my connections with the family of De Bourgh, and my relationship to your own, are circumstances highly in my favour; and you should take it into further consideration that, in spite of your manifold attractions, it is by no means certain that another offer of marriage may ever be made to you. Your portion is, unhappily, so small that it will in all likelihood undo the effects of your loveliness and amiable qualifications. As I must, therefore, conclude that you are not serious in your rejection of me, I shall choose to attribute it to your wish of increasing my love by suspense, according to the usual practice of elegant females.'

'I do assure you, sir, that I have no pretension whatever to that kind of elegance which consists in tormenting a respectable man. I would rather be paid the compliment of being believed sincere. I thank you again and again for the honour you have done me in your proposals, but to accept them is absolutely impossible. My feelings in every respect forbid it. Can I speak plainer? Do not consider me now as an elegant female intending to plague you, but as a rational creature speaking the truth from her heart.'

'You are uniformly charming!' cried he, with an air of awkward gallantry; 'and I am persuaded that, when sanctioned by the express authority of both your excellent parents, my proposals will not fail of being acceptable.'

To such perseverance in wilful self-deception Elizabeth would make no reply, and immediately and in silence withdrew, determined that, if he persisted in considering her repeated refusals as flattering encouragement, to apply to her father, whose negative might be uttered in such a manner as must be decisive, and whose behaviour at least could not be mistaken for the affectation and coquetry of an elegant female.

JANE AUSTEN, *Pride and Prejudice*, 1813

If any Lady not yet past her Grand Climacterick, of a Comfortable Fortune in her own Disposal, is desirous of spending the remainder of her Life with a tolerably handsome Fellow of great Parts, about five feet six inches, she may hear of such a one to her Mind by inquiring at the Theatre Coffee House for Mr F., a Sophister of Cambridge.

¶*An advertisement from an eighteenth-century newspaper.*

'I beg your pardon, miss!' said Mr Guppy, rising, when he saw me rise. 'But would you allow me the favour of a minute's private conversation?'

Not knowing what to say, I sat down again.

'What follows is without prejudice, miss?' said Mr Guppy, anxiously bringing a chair towards my table.

'I don't understand what you mean,' said I, wondering.

'It's one of our law terms, miss. You won't make any use of it to my detriment, at Kenge and Carboy's, or elsewhere. If our conversation shouldn't lead to anything, I am to be as I was, and am not to be prejudiced in my situation or worldly prospects. In short, it's in total confidence.'

'I am at a loss, sir,' said I, 'to imagine what you can have to communicate in total confidence to me, whom you have never seen but once; but I should be very sorry to do you any injury.'

'Thank you, miss. I'm sure of it—that's quite sufficient.' All this time Mr Guppy was either planing his forehead with his handkerchief, or tightly rubbing the palm of his left hand with the palm of his right. 'If you would excuse my taking another glass of wine, miss, I think it might assist me in getting on, without a continual choke that cannot fail to be mutually unpleasant.'

He did so, and came back again. I took the opportunity of moving well behind my table.

'You wouldn't allow me to offer you one, would you, miss?' said Mr Guppy, apparently refreshed.

'Not any,' said I.

'Not half a glass?' said Mr Guppy; 'quarter? No! Then, to proceed. My present salary, Miss Summerson, at Kenge and Carboy's, is two pound a week. When I first had the happiness of looking upon you, it was one-fifteen, and had stood at that figure for a lengthened period. A rise of five has since taken place, and a further rise of five is guaranteed at the expiration of a term not exceeding twelve months from the present date. My mother has a little property, which takes the form of a small life annuity; upon which she lives in an independent though unassuming manner, in the Old Street Road. She is eminently calculated for a mother-in-law. She never interferes, is all for peace, and her disposition easy. She has her failings—as who has not?—but I never knew her to do it when company was present; at which time you may freely trust her with wines, spirits, or malt liquors. My own abode is lodgings at Penton Place, Pentonville. It is lowly, but airy, open at the back, and considered one of the 'ealthiest outlets. Miss Summerson! In the mildest language, I adore you. Would you be so kind as to allow me (as I may say) to file a declaration—to make an offer!'

Mr Guppy went down on his knees. I was well behind my table, and not much frightened. I said, 'Get up from that ridiculous position immediately, sir, or you will oblige me to break my implied promise and ring the bell!'

'Hear me out, miss!' said Mr Guppy, folding his hands.

'I cannot consent to hear another word, sir,' I returned, 'unless you get up from the carpet directly, and go and sit down at the table, as you ought to do if you have any sense at all.'

He looked piteously, but slowly rose and did so.

'Yet what a mockery it is, miss,' he said, with his hand upon his heart, and shaking his head at me in a melancholy manner over the tray, 'to be stationed behind food at such a moment. The soul recoils from food at such a moment, miss.'

'I beg you to conclude,' said I; 'you have asked me to hear you out, and I beg you to conclude.'

'I will, miss,' said Mr Guppy. 'As I love and honour, so likewise I obey. Would that I could make Thee the subject of that vow, before the shrine!'

'That is quite impossible,' said I, 'and entirely out of the question.'

'I am aware,' said Mr Guppy, leaning forward over the tray and regarding me, as I again strangely felt, though my eyes were not directed to him, with his late intent look, 'I am aware that in a worldly point of view, according to all appearances, my offer is a poor one. But, Miss Summerson! Angel!—No, don't ring—I have been brought up in a sharp school, and am accustomed to a variety of general practice. Though a young man, I have ferreted out evidence, got up cases, and seen lots of life. Blest with your hand, what means might I not find of advancing your interests, and pushing your fortunes! What might I not get to know, nearly concerning you? I know nothing now, certainly; but what *might* I not, if I had your confidence, and you set me on?'

I told him that he addressed my interest, or what he supposed to be my interest, quite as unsuccessfully as he addressed my inclination; and he would now understand that I requested him, if he pleased, to go away immediately.

'Cruel miss,' said Mr Guppy, 'hear but another word! I think you must have seen that I was struck with those charms, on the day when I waited at the Whytorseller. I think you must have remarked that I could not forbear a tribute to those charms when I put up the steps of the 'ackney-coach. It was a feeble tribute to Thee, but it was well meant. Thy image has ever since been fixed in my breast. I have walked up and down, of an evening, opposite Jellyby's house, only to look upon the bricks that once contained Thee. This out of today, quite an unnecessary out so far as the attendance, which was its pretended object, went, was planned by me alone for Thee alone. If I speak of interest, it is only to recommend myself and my respectful wretchedness. Love was before it, and is before it.'

'I should be pained, Mr Guppy,' said I, rising and putting my hand upon the bell-rope, 'to do you, or any one who was sincere, the injustice of slighting any honest feeling, however disagreeably expressed. If you have really meant to give me a proof of your good opinion, though ill-timed and misplaced, I feel that I ought to thank you. I have very little reason to be proud, and I am not proud. I hope,' I think I added, without very well knowing what I said, 'that you will now go away as if you had never been so exceedingly foolish, and attend to Messrs Kenge and Carboy's business.'

'Half a minute, miss!' cried Mr Guppy, checking me as I was about to ring. 'This has been without prejudice?'

'I will never mention it,' said I, 'unless you should give me future occasion to do so.'

'A quarter of a minute, miss! In case you should think better—at any time, however distant, *that's* no consequence, for my feelings can never alter—of anything I have said, particularly what might I not do—Mr William Guppy, eighty-seven, Penton Place, or if removed, or dead (of blighted hopes or anything of that sort), care of Mrs Guppy, three hundred and two, Old Street Road, will be sufficient.'

I rang the bell, the servant came, and Mr Guppy, laying his written card upon the table, and making a dejected bow, departed.

CHARLES DICKENS, *Bleak House*, 1852–3

Whereas four young gentlemen, bachelors, in a pretty way of business, capable of rendering any four agreeable young ladies happy, lately disappointed in their amours, are resolved upon a matrimonial state by New Year's Day. If any ladies (Milliners excepted) have a mind to enter into the said state, let them enquire at the bar of Grigsby's Coffee House near the Royal Exchange between the hours of four and five for H.J., B.O., P.J., or C.J.

¶*An eighteenth-century advertisement. (Is 'milliner' a euphemism?)*

Tell you what you might do while you are alone at Pixton. You might think about me a bit & whether, if those wop priests ever come to a decent decision, you could bear the idea of marrying me. Of course you haven't got to decide, but think about it. I can't advise you in my favour because I think it would be beastly for you, but think how nice it would be for me. I am restless & moody & misanthropic & lazy & have no money except what I earn and if I got ill you would starve. In fact its a lousy proposition. On the other hand I think I could do a Grant and reform & become quite strict about not getting drunk and I am pretty sure I should be faithful. Also there is always a fair chance that there will be another bigger economic crash in which case if you had married a nobleman with a great house you might find yourself starving, while I am very clever and could probably earn a living of some sort somewhere. Also though you would be taking on an elderly buffer, I am one without fixed habits. You wouldn't find yourself confined to any particular place or group. Also I have practically no living relatives except one brother whom I scarcely know. You would not find yourself involved in a large family & all their rows & you would not be patronized & interfered with by odious sisters in law & aunts as often happens. All these are very small advantages compared with the awfulness of my character. I have always tried to be nice to you and you may have got it into your head that I am nice really, but that is all rot. It is only to you & for you. I am jealous & impatient—but there is no point in going into a whole list of my vices. You are a critical girl and I've no doubt that you know them all and a great many I don't know myself. But the point I wanted to make is that if you marry most people, you are marrying a great number of objects & other people as well, well if you marry me there is nothing else involved, and that is an advantage as well as a disadvantage. My only tie of any kind is my work. That means that for several months each year we shall have to separate or you would have to share some very lonely place with me. But apart from that we could do what we liked & go where we liked—and if you married a soldier or stockbroker or member of parliament or master of hounds you would be more tied. When I tell my friends that I am in love with a girl of 19 they looked shocked and say 'wretched child' but I dont look on you as very young even in your beauty and I dont think there is any sense in the line that you cannot possibly commit yourself to a decision that affects your whole life for years yet. But anyway there is no point in your deciding or even answering. I may never get free of your cousin Evelyn. Above all things, darling, dont fret at all. But just turn the matter over in your dear head. . . .

Eight days from now I shall be with you again, darling heart. I don't think of much else.

All my love
Evelyn

¶ *from a letter from Evelyn Waugh to Laura Herbert, whom he married in 1937, after his marriage to her cousin Evelyn had been annulled.*

Lætitia . . . would not descend to the family breakfast-table. Clara would fain have stayed to drink tea with her in her own room, but a last act of conformity was demanded of the liberated young lady. She promised to run up the moment breakfast was over. Not unnaturally, therefore, Lætitia supposed it to be she to whom she gave admission, half an hour later, with a glad cry of, 'Come in, dear.'

The knock had sounded like Clara's.

Sir Willoughby entered.

He stepped forward. He seized her hands. 'Dear!' he said. 'You cannot withdraw that. You called me dear. I am, I must be dear to you. The word is out, by accident or not, but, by heaven, I have it and I give it up to no one. And love me or not—marry me, and my love will bring it back to you. You have taught me I am not so strong. I must have you by my side. You have powers I did not credit you with.'

'You are mistaken in me, Sir Willoughby,' Lætitia said feebly, outworn as she was.

'A woman who can resist me by declining to be my wife, through a whole night of entreaty, has the quality I need for my house, and I batter at her ears for months, with as little rest as I had last night, before I surrender my chance of her. But I told you last night I want you within the twelve hours. I have staked my pride on it. By noon you are mine: you are introduced to Mrs Mountstuart as mine, as the lady of my life and house. And to the world! I shall not let you go.'

'You will not detain me here, Sir Willoughby?'

'I will detain you. I will use force and guile. I will spare nothing.'

He raved for a term, as he had done overnight.

On his growing rather breathless, Lætitia said: 'You do not ask me for love?'

'I do not. I pay you the higher compliment of asking for *you*, love or no love. My love shall be enough. Reward me or not. I am not used to be denied.'

'But do you know what you ask for? Do you remember what I told you of myself? I am hard, materialistic; I have lost faith in romance, the skeleton is present with me all over life. And my health is not good. I crave for money. I should marry to be rich. I should not worship you. I should be a burden, barely a living one, irresponsive and cold. Conceive such a wife, Sir Willoughby!'

'It will be you!'

She tried to recall how this would have sung in her ears long back. Her bosom rose and fell in absolute dejection. Her ammunition of arguments against him had been expended overnight.

'You are so unforgiving,' she said.

'Is it I who am?'

'You do not know me.'

'But you are the woman of all the world who knows *me*, Lætitia.'

'Can you think it better for you to be known?'

He was about to say other words: he checked them. 'I believe I do not know myself. Anything you will, only give me your hand; give it; trust to me; you shall direct me. If I have faults, help me to obliterate them.'

'Will you not expect me to regard them as the virtues of meaner men?'

'You will be my wife!'

Lætitia broke from him, crying: 'Your wife, your critic! Oh!

I cannot think it possible. Send for the ladies. Let them hear me.'

'They are at hand,' said Willoughby, opening the door.

They were in one of the upper rooms anxiously on the watch.

'Dear ladies,' Lætitia said to them, as they entered. 'I am going to wound you, and I grieve to do it: but rather now than later, if I am to be your housemate. He asks me for a hand that cannot carry a heart, because mine is dead. I repeat it. I used to think the heart a woman's marriage portion for her husband. I see now that she may consent, and he accept her, without one. But it is right that you should know what I am when I consent. I was once a foolish romantic girl; now I am a sickly woman, all illusions vanished. Privation has made me what an abounding fortune usually makes of others—I am an Egoist. I am not deceiving you. That is my real character. My girl's view of him has entirely changed; and I am almost indifferent to the change. I can endeavour to respect him, I cannot venerate.'

'Dear child!' the ladies gently remonstrated.

Willoughby motioned to them.

'If we are to live together, and I could very happily live with you,' Lætitia continued to address them, 'you must not be ignorant of me. And if you, as I imagine, worship him blindly, I do not know how we are to live together. And never shall you quit this house to make way for me. I have a hard detective eye. I see many faults.'

'Have we not all of us faults, dear child?'

'Not such as he has; though the excuses of a gentleman nurtured in idolatry may be pleaded. But he should know that they are seen, and seen by her he asks to be his wife, that no misunderstanding may exist, and while it is yet time he may consult his feelings. He worships himself.'

'Willoughby?'

'He is vindictive.'

'Our Willoughby?'

'That is not your opinion, ladies. It is firmly mine.'

*

'Ladies, you are witnesses that there is no concealment, there has been no reserve, on my part. May heaven grant me kinder eyes than I have now. I would not have you change your opinion of him; only that you should see how I read him. For the rest, I vow to do my duty by him. Whatever is of worth in me is at his service. I am very tired. I feel I must yield or break. This is his wish, and I submit.'

'And I salute my wife,' said Willoughby, making her hand his own, and warming to his possession as he performed the act.

GEORGE MEREDITH, *The Egoist*, 1879

Vivacious 42-year-old divorcee seeks male companion with similar personality for evenings out, must have own teeth.

¶ *An advertisement from* The Lakeland Echo, *1985*.

[*Enter Millamant and Mirabell*]

MIRABELL: *Like Daphne she, as lovely and as coy.* Do you lock yourself up from me, to make my search more curious? or is this pretty artifice contrived to signify that here the chase must end, and my pursuits be crowned? For you can fly no further.

MILLAMANT: Vanity! no—I'll fly, and be followed to the last moment. Though I am upon the very verge of matrimony, I expect you should solicit me as much as if I were wavering at the gate of a monastery, with one foot over the threshold. I'll be solicited to the very last, nay, and afterwards.

MIRABELL: What, after the last?

MILLAMANT: Oh, I should think I was poor and had nothing to bestow, if I were reduced to an inglorious ease, and freed from the agreeable fatigues of solicitation.

MIRABELL: But do not you know, that when favours are conferred upon instant and tedious solicitation, that they diminish in their value, and that both the giver loses the grace, and the receiver lessens his pleasure?

MILLAMANT: It may be in things of common application; but never sure in love. Oh, I hate a lover that can dare to think he draws a moment's air, independent on the bounty of his mistress. There is not so impudent a thing in nature, as the saucy look of an assured man, confident of success. The pedantic arrogance of a very husband has not so pragmatical an air. Ah! I'll never marry, unless I am first made sure of my will and pleasure.

MIRABELL: Would you have 'em both before marriage? or will you be contented with the first now, and stay for the other till after grace?

MILLAMANT: Ah! don't be impertinent.—My dear liberty, shall I leave thee? my faithful solitude, my darling contemplation, must I bid you then adieu? Ay-h adieu—my morning thoughts, agreeable wakings, indolent slumbers, all ye *douceurs*, ye *sommeils du matin*, adieu?—I can't do't, 'tis more than impossible—positively, Mirabell, I'll lie abed in a morning as long as I please.

MIRABELL: Then I'll get up in a morning as early as I please.

MILLAMANT: Ah! idle creature, get up when you will—and d'ye hear, I won't be called names after I'm married; positively I won't be called names.

MIRABELL: Names!

MILLAMANT: Ay, as wife, spouse, my dear, joy, jewel, love, sweetheart, and the rest of that nauseous cant, in which men and their wives are so fulsomely familiar—I shall never bear that—good Mirabell, don't let us be familiar or fond, nor kiss before folks, like my lady Fadler, and sir Francis: nor go to Hyde Park together the first Sunday in a new chariot, to provoke eyes and whispers, and then never to be seen there together again; as if we were proud of one another the first week, and ashamed of one another ever after. Let us never visit together, nor go to a play together; but let us be very strange and well bred: let us be as strange as if we had been married a great while; and as well bred as if we were not married at all.

MIRABELL: Have you any more conditions to offer? Hitherto your demands are pretty reasonable.

MILLAMANT: Trifles!—As liberty to pay and receive visits to and from whom I please; to write and receive letters, without interrogatories or wry faces on your part; to wear what I please; and choose conversation with regard only to my own taste; to have no obligation upon me to converse with wits that I don't like, because they are your acquaintance; or to be intimate with fools, because they may be your relations. Come to dinner when I please; dine in my dressing-room when I'm out of humour, without giving a reason. To have my closet inviolate; to be sole empress of my tea-table, which you must never presume to approach without first asking leave. And, lastly, wherever I am, you shall always knock at the door before you come in. These articles subscribed, if I continue to endure you a little longer, I may by degrees dwindle into a wife.

MIRABELL: Your bill of fare is something advanced in this latter account.—Well, have I liberty to offer conditions—that when you are dwindled into a wife, I may not be beyond measure enlarged into a husband?

MILLAMANT: You have free leave; propose your utmost, speak and spare not.

MIRABELL: I thank you.—*Imprimis* then, I covenant, that your acquaintance be general; that you admit no sworn confidant, or intimate of your own sex; no she friend to screen her affairs under your countenance, and tempt you to make trial of a mutual secrecy. No decoy duck to wheedle you a fop-scrambling to the play in a mask—then bring you home in a pretended fright, when you think you shall be found out—and rail at me for missing the play, and disappointing the frolic which you had to pick me up, and prove my constancy.

MILLAMANT: Detestable *imprimis*! I go to the play in a mask!

MIRABELL: *Item*, I article, that you continue to like your own face, as long as I shall: and while it passes current with me, that you endeavour not to new-coin it. To which end, together with all vizards for the day, I prohibit all masks for the night, made of oiled-skins, and I know not what—hogs' bones, hares' gall, pig-water, and the marrow of a roasted cat. In short, I forbid all commerce with the gentlewoman in what d'ye call it court. *Item*, I shut my doors against all bawds with baskets, and penny-worths of muslin, china, fans, atlasses, etc.—*Item*, when you shall be breeding——

MILLAMANT: Ah! name it not.

MIRABELL: Which may be presumed with a blessing on our endeavours——

MILLAMANT: Odious endeavours!

MIRABELL: I denounce against all strait lacing, squeezing for a shape, till you mould my boy's head like a sugar-loaf, and instead of a man-child, make me father to a crooked billet. Lastly, to the dominion of the tea-table I submit—but with proviso, that you exceed not in your province; but restrain yourself to native and simple tea-table drinks, as tea, chocolate, and coffee: as likewise to genuine and authorized tea-table talk—such as mending of fashions, spoiling reputations, railing at absent friends, and so forth—but that on no account you encroach upon the men's prerogative, and presume to drink healths, or toast fellows; for prevention of which I banish all foreign forces, all auxiliaries to the tea-table, as orange-brandy, all aniseed, cinnamon, citron and

Barbadoes-waters, together with ratafia, and the most noble spirit of clary—but for cowslip wine, poppy water, and all dormitives, those I allow.—These provisos admitted, in other things I may prove a tractable and complying husband.

MILLAMANT: O horrid provisos! filthy strong-waters! I toast fellows! odious men! I hate your odious provisos.

MIRABELL: Then we are agreed! shall I kiss your hand upon the contract? And here comes one to be a witness to the sealing of the deed.

WILLIAM CONGREVE, *The Way of the World*, 1700

The Vicar proposes to the Governess:

'May I come to what is in my mind?'

'I think you had better not. There is a story about someone who saw into people's minds, and it was impossible for him. And what is the good of not being able to see into them, if you are told about it?'

'You would have to be told. You can have no inkling. It would transcend your furthest dreams. And if you could guess it, your tongue would be barred.' Chaucer spoke with a great gentleness of a woman's further compulsions. 'I will go slowly. You shall have time.'

'What can it be? I thought you were going to propose to me. But that has nothing to do with my dreams. Have I not seen into your mind after all?'

Chaucer looked at her in silence.

'You forgot a woman's intuition, when you enumerated the things about her. And you forgot her tongue. It is not so often barred. It is not really supposed to be. Is it going to be yours that is barred?'

'Did you want me to say it in words?' said Chaucer, struck by a thought.

'No, no, your way of saying it was much better. Your tongue was so nicely barred. I hope it will continue to be. You see how mistaken you were in me. Suppose I were to accept you now?'

'I should be honoured.'

'You would be, I suppose, more than you would have been. But it is no good to marry a dependant, if she is not grateful. You might as well marry someone who had no need to be.'

'Am I to understand you do not wish for what I offer?'

'Well, don't you understand it? My tongue has not been barred. I hope yours is not going to break its bars.'

'It is not,' said Chaucer, bowing. 'I hope I know how to conduct myself under a refusal. I have not much to offer a woman, but some might have found it acceptable.'

'It was to transcend my furthest dreams. So no wonder you thought I should find it so. I quite understand it.'

There was a pause, and then Chaucer spoke.

'Well, perhaps you know what is best for us. I will take your decision as the right one.'

'Will you really? I did not expect you to do that. I did not know that men who were refused, took the decision as the right one. I think it is I who have been refused. It has to be made to appear that it is the man. Perhaps it is really I who have proposed?'

IVY COMPTON-BURNETT, *Daughters and Sons*, 1937

They proceeded to the drawing-room. . . . One could visualize the ladies withdrawing to it, while their lords discussed life's realities below, to the accompaniment of cigars. . . .

Just as this thought entered Margaret's brain, Mr Wilcox did ask her to be his wife, and the knowledge that she had been right so overcame her that she nearly fainted.

But the proposal was not to rank among the world's great love scenes.

'Miss Schlegel'—his voice was firm—'I have had you up on false pretences. I want to speak about a much more serious matter than a house.'

Margaret almost answered: 'I know—'

'Could you be induced to share my—is it probable—'

'Oh, Mr Wilcox!' she interrupted, holding the piano and averting her eyes. 'I see, I see. I will write to you afterwards if I may.'

He began to stammer. 'Miss Schlegel—Margaret—you don't understand.'

'Oh, yes! Indeed, yes!' said Margaret.

'I am asking you to be my wife.'

So deep already was her sympathy, that when he said, 'I am asking you to be my wife,' she made herself give a little start. She must show surprise if he expected it. An immense joy came over her. It was indescribable. It had nothing to do with humanity, and most resembled the all-pervading happiness of fine weather. Fine weather is due to the sun, but Margaret could think of no central radiance here. She stood in his drawing-room happy, and longing to give happiness. On leaving him she realized that the central radiance had been love.

'You aren't offended, Miss Schlegel?'

'How could I be offended?'

There was a moment's pause. He was anxious to get rid of her, and she knew it. She had too much intuition to look at him as he struggled for possessions that money cannot buy. He desired comradeship and affection, but he feared them, and she, who had taught herself only to desire, and could have clothed the struggle with beauty, held back, and hesitated with him.

'Good-bye,' she continued. 'You will have a letter from me—I am going back to Swanage to-morrow.'

'Thank you.'

'Good-bye, and it's you I thank.'

'I may order the motor round, mayn't I?'

'That would be most kind.'

'I wish I had written instead. Ought I to have written?'

'Not at all.'

'There's just one question—'

She shook her head. He looked a little bewildered, and they parted.

They parted without shaking hands: she had kept the interview, for his sake, in tints of the quietest grey. Yet she thrilled with happiness ere she reached her own house. Others had loved her in the past, if one may apply to their brief desires so grave a word, but those others had been 'ninnies'—young men who had nothing to do, old men who could find nobody better. And she had often 'loved', too, but only so far as the facts of sex demanded: mere yearnings for the masculine, to be dismissed for what they were worth, with a smile. Never before had her personality been touched. She was not young or very rich, and it amazed her that a man of any standing should take her seriously. As she sat trying to do accounts in her empty house, amidst beautiful pictures and noble books, waves of emotion broke, as if a tide of passion was flowing through the night air. She shook her head, tried to concentrate her attention, and failed. In vain did she repeat: 'But I've been through this sort of thing before.' She had never been through it; the big machinery, as opposed to the little, had been set in motion, and the idea that Mr Wilcox loved, obsessed her before she came to love him in return.

She would come to no decision yet. 'Oh, sir, this is so sudden'—that prudish phrase exactly expressed her when her time came.

Premonitions are not preparations. She must examine more closely her own nature and his; she must talk it over judicially with Helen. It had been a strange love-scene—the central radiance unacknowledged from first to last. She, in his place, would have said 'Ich liebe dich', but perhaps it was not his habit to open the heart. He might have done it if she had pressed him—as a matter of duty, perhaps; England expects every man to open his heart once; but the effort would have jarred him, and never, if she could avoid it, should he lose those defences that he had chosen to raise against the world. He must never be bothered with emotional talk, or with a display of sympathy. He was an elderly man now, and it would be futile and impudent to correct him.

E. M. FORSTER, *Howards End*, 1910

Benjamin Franklin came to France as American Ambassador in 1776. Here, aged 72 and widowed for six years, he fell in love with and proposed to the widow of the Swiss philosopher Helvétius, who was herself aged 61. He wrote her the following letter after she had refused his offer of marriage:

Passy [January, 1780]

Chagrined at your resolution, pronounced so decidedly last evening, to remain single for life, in honour of your dear husband, I went home, fell upon my bed, thought myself dead, and found myself in the Elysian Fields.

They asked me if I had any desire to see any persons in particular. 'Lead me to the philosophers.' 'There are two that reside here in this garden. They are very good neighbours and very friendly to each other.' 'Who are they?' 'Socrates and Helvétius.' 'I esteem them both prodigiously; but let me see Helvétius first, because I understand a little French and not a word of Greek.' He viewed me with much courtesy, having known me, he said, by reputation for some time. He asked me a thousand things about the war, and the present state of religion, liberty and government in France. 'You ask me nothing, then, respecting your friend Madame Helvétius, and yet she loves you still excessively; it is but an hour since I was at her house.' 'Ah!' said he, 'you make me recollect my former felicity; but I ought to forget it to be happy here. For many years I thought of nothing but her. At last I am consoled. I have taken another wife, the most like her that I could find. She is not, it is true, quite so handsome; but she has as much good sense and wit, and loves me infinitely. Her continued study is to please me; she is at present gone to look for the best nectar and ambrosia to regale me this evening; stay with me and you will see her.'

'I perceive,' said I, 'that your old friend is more faithful than you; for many good matches have been offered her, all of which she has refused. I confess to you that I loved her myself to excess; but she was so severe to me, and has absolutely refused me, for love of you.' 'I commiserate you,' said he, 'for your misfortune; for indeed she is a good woman, and very amiable. But the Abbé de la Roche and the Abbé Morellet, are they not still sometimes at her house?' 'Yes, indeed, for she has not lost a single one of your friends.' 'If you had gained over the Abbé Morellet with coffee and cream to speak for you, perhaps you would have succeeded, for he is as subtle a reasoner as Scotus or St Thomas, and puts his arguments in such good order that they become almost irresistible: or if you had secured the Abbé de la Roche, by giving him some fine edition of an old classic, to speak *against* you, that would have been better; for I have always observed that when he advises anything, she has a strong inclination to do the reverse.'

At these words the new Madame Helvétius entered with the nectar; I instantly recognised her as Mrs Franklin, my old American friend. I reclaimed her, but she said to me coldly, 'I have been your good wife forty-nine years and four months; almost half a century; be content with that.' Dissatisfied with this refusal of my Eurydice I immediately resolved to quit those ungrateful shades and to return to this good world to see again the sun and you. Here I am. Let us avenge ourselves.

BENJAMIN FRANKLIN

¶ *Mme Helvétius continued to refuse Franklin, but they remained very close friends.*

KING HENRY: Fair Katharine, and most fair,
Will you vouchsafe to teach a soldier terms
Such as will enter at a lady's ear
And plead his love-suit to her gentle heart?

KATHARINE: Your majesty shall mock at me; I cannot speak your England.

KING HENRY: O fair Katharine, if you will love me soundly with your French heart, I will be glad to hear you confess it brokenly with your English tongue. Do you like me, Kate?

KATHARINE: Pardonnez-moi, I cannot tell vat is 'like me.'

KING HENRY: An angel is like you, Kate, and you are like an angel.

KATHARINE: Que dit-il? que je suis semblable à les anges?

ALICE: Oui, vraiment, sauf votre grace, ainsi dit-il.

KING HENRY: I said so, dear Katharine; and I must not blush to affirm it.

KATHARINE: O bon Dieu! les langues des hommes sont pleines de tromperies.

KING HENRY: What says she, fair one? that the tongues of men are full of deceits?

ALICE: Oui, dat de tongues of de mans is be full of deceits: dat is de princess.

KING HENRY: The princess is the better Englishwoman. I' faith, Kate, my wooing is fit for thy understanding: I am glad thou canst speak no better English; for, if thou couldst, thou wouldst find me such a plain king that thou wouldst think I had sold my farm to buy my crown. I know no ways to mince it in love, but directly to say 'I love you:' then if you urge me farther than to say 'Do you in faith?' I wear out my suit. Give me your answer; i' faith, do: and so clap hands and a bargain: how say you, lady?

KATHARINE: Sauf votre honneur, me understand vell.

KING HENRY: Marry, if you would put me to verses or to dance for your sake, Kate, why you undid me; for the one, I have neither words nor measure, and for the other, I have no strength in measure, yet a reasonable measure in strength. If I could win a lady at leap-frog, or by vaulting into my saddle with my armour on my back, under the correction of bragging be it spoken, I should quickly leap into a wife. Or if I might buffet for my love, or bound my horse for her favours, I could lay on like a butcher and sit like a jack-an-apes, never off. But, before God, Kate, I cannot look greenly nor gasp out my eloquence, nor I have no cunning in protestation; only downright oaths, which I never use till urged, nor never break for urging. If thou canst love a fellow of this temper, Kate, whose face is not worth sun-burning, that never looks in his glass for love of any thing he sees there, let thine eye be thy cook. I speak to thee plain soldier: if thou canst love me for this, take me; if not, to say to thee that I shall die, is true; but for thy love, by the Lord, no; yet I love thee too. And while thou livest, dear Kate, take a fellow of plain and uncoined constancy; for he perforce must do thee right, because he hath not the gift to woo in other places: for these fellows of infinite tongue, that can rhyme themselves into ladies' favours, they do always reason themselves out again. What! a speaker is but a prater; a rhyme is but a ballad. A good leg will fall; a straight back will stoop; a black beard will turn white; a curled pate will grow bald; a fair face will wither; a full eye will wax hollow: but a good heart, Kate, is the sun and the moon; or rather, the sun, and not the moon; for it shines bright and never changes, but keeps his course truly. If thou would have such a one, take me; and take me, take a soldier; take a soldier, take a king. And what sayest thou then to my love? speak, my fair, and fairly, I pray thee.

KATHARINE: Is it possible dat I sould love de enemy of France?

KING HENRY: No; it is not possible you should love the enemy of France, Kate: but, in loving me, you should love the friend of France; for I love France so well that I will not part with a village of it; I will have it all mine: and, Kate, when France is mine and I am yours, then yours is France and you are mine.

KATHARINE: I cannot tell vat is dat.

KING HENRY: No, Kate? I will tell thee in French; which I am sure will hang upon my tongue like a new-married wife about her husband's neck, hardly to be shook off. Je quand sur le possession de France, et quand vous avez le possession de moi,—let me see, what then? Saint Denis be my speed!—donc votre est France et vous êtes mienne. It is as easy for me, Kate, to conquer the kingdom as to speak so much more French: I shall never move thee in French, unless it be to laugh at me.

KATHARINE: Sauf votre honneur, le François que vous parlez, il est meilleur que l'Anglois lequel je parle.

KING HENRY: No, faith, is't not, Kate: but thy speaking of my tongue, and I thine, most truly-falsely, must needs be granted to be much at one. But, Kate, dost thou understand thus much English, canst thou love me?

KATHARINE: I cannot tell.

*

KING HENRY: Come, your answer in broken music; for thy voice is music and thy English broken; therefore, queen of all, Katharine, break thy mind to me in broken English, wilt thou have me?

KATHARINE: Dat is as it sall please de roi mon père.

KING HENRY: Nay, it will please him well, Kate; it shall please him, Kate.

KATHARINE: Den it sall also content me.

KING HENRY: Upon that I kiss your hand, and I call you my queen.

KATHARINE: Laissez, mon seigneur, laissez, laissez: ma foi, je ne veux point que vous abaissiez votre grandeur en baisant la main d'une de votre seigneurie indigne serviteur; excusez-moi, je vous supplie, mon très-puissant seigneur.

KING HENRY: Then I will kiss your lips, Kate.

KATHARINE: Les dames et demoiselles pour être baisées devant leur noces, il n'est pas la coutume de France.

KING HENRY: Madam my interpreter, what says she?

ALICE: Dat it is not be de fashion pour les ladies of France,—I cannot tell vat is baiser en Anglish.

KING HENRY: To kiss.

ALICE: Your majesty entendre bettre que moi.

KING HENRY: It is not a fashion for the maids in France to kiss before they are married, would she say?

ALICE: Oui, vraiment.

KING HENRY: O Kate, nice customs courtesy to great kings. Dear Kate, you and I cannot be confined within the weak list of a country's fashion: we are the makers of manners, Kate; and the liberty that follows our places stops the mouths of all find-faults; as I will do yours, for upholding the nice fashion of your country in denying me a kiss: therefore, patiently and yielding. [*Kissing her.*] You have witchcraft in your lips, Kate: there is more eloquence in a sugar touch of them than in the tongues of the French council; and they should sooner persuade Harry of England than a general petition of monarchs. Here comes your father.

WILLIAM SHAKESPEARE, *Henry V*, 1600

14 October 1839

After a little pause, I said to Lord M[elbourne], that I had made up my mind (about marrying dearest Albert). . . . 'I think it'll be very well received; for I hear there is an anxiety now that it should be; and I'm very glad of it; I think it is a very good thing, and you'll be much more comfortable; for a woman cannot stand alone for long, in whatever situation she is.' . . . Then I asked, if I hadn't better tell Albert of my decision soon, in which Lord M. agreed. How? I asked, for that in general such things were done the other way,—which made Lord M. laugh.

15 October 1839

At about ½ p. 12 I sent for Albert; he came to the Closet where I was alone, and after a few minutes I said to him, that I thought he must be aware why I wished [him] to come here, and that it would make me too happy if he would consent to what I wished (to marry me); we embraced each other over and over again, and he was so kind, so affectionate; Oh! to feel I was, and am, loved by such an Angel as Albert was too great delight to describe! he is perfection; perfection in every way—in beauty—in everything! I told him I was quite unworthy of him and kissed his dear hand—he said he would be very happy [to share his life with her] and was so kind and seemed so happy, that I really felt it was the happiest brightest moment in my life, which made up for all I had suffered and endured. Oh! how I adore and love him, I cannot say!! how I will strive to make him feel as little as possible the great sacrifice he has made; I told him it was a great sacrifice,—which he wouldn't allow . . . I feel the happiest of human beings.

QUEEN VICTORIA, from her Journal

MIRANDA: I do not know
One of my sex; no woman's face remember,
Save, from my glass, mine own; nor have I seen
More that I may call men than you, good friend,
And my dear father: how features are abroad,
I am skilless of; but, by my modesty,
The jewel in my dower, I would not wish
Any companion in the world but you;
Nor can imagination form a shape,
Besides yourself, to like of. But I prattle
Something too wildly, and my father's precepts
I therein do forget.

FERDINAND: I am, in my condition,
A prince, Miranda; I do think, a king;
I would not so!—and would no more endure

This wooden slavery than to suffer
The flesh-fly blow my mouth. Hear my soul speak:
The very instant that I saw you, did
My heart fly to your service; there resides,
To make me slave to it; and for your sake
Am I this patient log-man.

MIRANDA: Do you love me?

FERDINAND: O heaven, O earth, bear witness to this sound,
And crown what I profess with kind event
If I speak true! if hollowly, invert
What best is boded me to mischief! I,
Beyond all limit of what else i' the world,
Do love, prize, honour you.

MIRANDA: I am a fool
To weep at what I am glad of.

PROSPERO: Fair encounter
Of two most rare affections! Heavens rain grace
On that which breeds between 'em!

FERDINAND: Wherefore weep you?

MIRANDA: At mine unworthiness, that dare not offer
What I desire to give; and much less take
What I shall die to want. But this is trifling;
And all the more it seeks to hide itself
The bigger bulk it shows. Hence, bashful cunning!
And prompt me, plain and holy innocence!
I am your wife, if you will marry me;
If not, I'll die your maid: to be your fellow
You may deny me; but I'll be your servant,
Whether you will or no.

FERDINAND: My mistress, dearest;
And I thus humble ever.

MIRANDA: My husband, then?

FERDINAND: Ay, with a heart as willing
As bondage e'er of freedom: here's my hand.

MIRANDA: And mine, with my heart in't: and now farewell
Till half an hour hence.

FERDINAND: A thousand thousand!

WILLIAM SHAKESPEARE, *The Tempest*, 1611

She saw him the instant he entered the room. She had been waiting for him. She was filled with joy, and so confused at her own joy, that there was a moment—the moment when he went up to his hostess and glanced again at her—when she, and he, and Dolly, who saw it all, thought she would break down and burst into tears. Blushing and going pale by turns, she sat rigid, waiting with quivering lips for his approach. He went up to her, bowed, and silently held out his hand. Except for the slight quiver of her lips and the moist film that came over her eyes, making them appear brighter, her smile was almost calm as she said:

'What a long time it is since we met!' and with desperate resolve her cold hand pressed his.

*

They resumed the conversation started at dinner—the emancipation and occupations of women. Levin agreed with Dolly that a girl who did not marry could always find some feminine occupation in the family. He supported this view by saying that no family can get along without women to help them, that every family, poor or rich, had to have nurses, either paid or belonging to the family.

'No,' said Kitty, blushing, but looking at him all the more boldly with her truthful eyes, 'a girl may be so placed that it is humiliating for her to live in the family, while she herself. . .'

He understood her allusion.

'Oh yes,' he said. 'Yes, yes, yes—you're right; you're right!'

And he saw all that Pestsov had been driving at at dinner about the freedom of women, simply because he got a glimpse of the terror in Kitty's heart of the humiliation of remaining an old maid; and, loving her, he felt that terror and humiliation, and at once gave up his contention.

A silence followed. She continued scribbling on the table with the chalk. Her eyes shone with a soft light. Surrendering to her mood he felt a continually growing tension of happiness throughout his whole being.

'Oh, I've scribbled all over the table!' she exclaimed, and, putting down the chalk, made a movement to get up.

'What! Shall I be left alone—without her?' he thought, with terror, and took the piece of chalk. 'Don't go,' he said, sitting down at the table. 'I've wanted to ask you a question for a long time.' He looked straight into her caressing, though frightened eyes.

'What is it?'

'Here,' he said, and wrote down the initial letters, w, y, t, m, *i*, *c*, *n*, *b*—d, t, m, n, o, t? These letters stood for, 'When you told me *it could not be*—did that mean never, or then?' There seemed no likelihood that she would be able to decipher this complicated sequence; but he looked at her as though his life depended on her understanding the words.

She gazed up at him seriously, then leaned her puckered forehead on her hand and began to read. Once or twice she stole a look at him, as though asking, 'Is it what I think?'

'I know what it is,' she said, flushing a little.

'What is this word?' he asked, pointing to the *n* which stood for *never*.

'That means *never*,' she said, 'but it's not true!'

He quickly rubbed out what he had written, handed her the chalk, and stood up. She wrote: *T*, *I*, *c*, *n*, *a*, *d*.

Dolly felt consoled for the grief caused by her conversation with Karenin when she caught sight of the two together: Kitty with the chalk in her hand, gazing up at Levin with a shy, happy smile, and his fine figure bending over the table, his radiant eyes directed now on the table, now on her. He was suddenly radiant: he had understood. The letters meant: 'Then I could not answer differently.'

He glanced at her questioningly, timidly.

'Only then?'

'Yes,' her smile answered.

'And n . . .—and now?' he asked.

'Well, read this. I'll tell you what I should like, what I should like so much!' She wrote the initial letters: *i*, *y*, *c*, *f*, *a*, *f*, *w*, *h*, meaning, 'If you could forget and forgive what happened.'

He seized the chalk, breaking it with his nervous, trembling fingers, and wrote the first letters of the following sentence: 'I have nothing to forget and forgive; I have never ceased to love you.'

She looked at him with a smile that did not waver.

'I understand,' she said in a whisper.

He sat down and wrote a long sentence. She understood it all and, without asking if she was right, took the chalk and at once wrote the answer.

For a long time he could not make out what it was, and kept looking up into her eyes. He was dazed with happiness. He could not fill in the words she meant at all; but in her lovely eyes, suffused with happiness, he saw all he needed to know. And he wrote down three letters. But before he had finished writing she read them over his arm, and herself finished and wrote the answer, 'Yes.'

LEO TOLSTOY, *Anna Karenina*, 1877

At this point Kate ceased to attend. He saw after a little that she had been following some thought of her own, and he had been feeling the growth of something determinant even through the extravagance of much of the pleasantry, the warm transparent irony, into which their livelier intimacy kept plunging like a confident swimmer. Suddenly she said to him with extraordinary beauty: 'I engage myself to you for ever.'

The beauty was in everything, and he could have separated nothing—couldn't have thought of her face as distinct from the whole joy. Yet her face had a new light. 'And I pledge you—I call God to witness!—every spark of my faith; I give you every drop of my life.' That was all, for the moment, but it was enough, and it was almost as quiet as if it were nothing. They were in the open air, in an alley of the Gardens; the great space, which seemed to arch just then higher and spread wider for them, threw them back into deep concentration. They moved by a common instinct to a spot, within sight, that struck them as fairly sequestered, and there, before their time together was spent, they had extorted from concentration every advance it could make them. They had exchanged vows and tokens, sealed their rich compact, solemnised, so far as breathed words and murmured sounds and lighted eyes and clasped hands could do it, their agreement to belong only, and to belong tremendously, to each other.

HENRY JAMES, *The Wings of the Dove*, 1902

'My dear Louisa,' said her father, 'I prepared you last night to give me your serious attention in the conversation we are now going to have together. You have been so well trained, and you do, I am happy to say, so much justice to the education you have received, that I have perfect confidence in your good sense. You are not impulsive, you are not romantic, you are accustomed to view everything from the strong dispassionate ground of reason and calculation. From that ground alone, I know you will view and consider what I am going to communicate.'

He waited, as if he would have been glad that she said something. But she said never a word.

'Louisa, my dear, you are the subject of a proposal of marriage that has been made to me.'

Again he waited, and again she answered not one word. This so far surprised him, as to induce him gently to repeat, 'a proposal of marriage, my dear.' To which she returned, without any visible emotion whatever:

'I hear you, father. I am attending, I assure you.'

'Well!' said Mr Gradgrind, breaking into a smile, after being for the moment at a loss, 'you are even more dispassionate than I expected, Louisa. Or, perhaps, you are not unprepared for the announcement I have it in charge to make?'

'I cannot say that, father, until I hear it. Prepared or unprepared, I wish to hear it all from you. I wish to hear you state it to me, father.'

Strange to relate, Mr Gradgrind was not so collected at this moment as his daughter was. He took a paper-knife in his hand, turned it over, laid it down, took it up again, and even then had to look along the blade of it, considering how to go on.

'What you say, my dear Louisa, is perfectly reasonable. I have undertaken then to let you know that—in short, that Mr Bounderby has informed me that he has long watched your progress with particular interest and pleasure, and has long hoped that the time might ultimately arrive when he should offer you his hand in marriage. That time, to which he has so long, and certainly with great constancy, looked forward, is now come. Mr Bounderby has made his proposal of marriage to me, and has entreated me to make it known to you, and to express his hope that you will take it into your favourable consideration.'

Silence between them. The deadly statistical clock very hollow. The distant smoke very black and heavy.

'Father,' said Louisa, 'do you think I love Mr Bounderby?'

Mr Gradgrind was extremely discomfited by this unexpected question. 'Well, my child,' he returned, 'I—really—cannot take upon myself to say.'

'Father,' pursued Louisa in exactly the same voice as before, 'do you ask me to love Mr Bounderby?'

'My dear Louisa, no. No. I ask nothing.'

'Father,' she still pursued, 'does Mr Bounderby ask me to love him?'

'Really, my dear,' said Mr Gradgrind, 'it is difficult to answer your question——'

'Difficult to answer it, Yes or No, father?'

'Certainly, my dear. Because;' here was something to demonstrate, and it set him up again; 'because the reply depends so materially, Louisa, on the sense in which we use the expression. Now, Mr Bounderby does not do you the injustice, and does not do himself the injustice, of pretending to anything fanciful, fantastic, or (I am using synonymous terms) sentimental. Mr Bounderby would have seen you grow up under his eyes, to very little purpose, if he could so far forget what is due to your good sense, not to say to his, as to address you from any such ground. Therefore, perhaps the expression itself—I merely suggest this to you, my dear—may be a little misplaced.'

*

'What do you recommend, father,' asked Louisa, . . . 'that I should substitute for the term I used just now? For the misplaced expression?'

'Louisa,' returned her father, 'it appears to me that nothing can be plainer. Confining yourself rigidly to Fact, the question of Fact you state to yourself is: Does Mr Bounderby ask me to marry him? Yes, he does. The sole remaining question then is: Shall I marry him? I think nothing can be plainer than that?'

'Shall I marry him?' repeated Louisa, with great deliberation.

'Precisely. And it is satisfactory to me, as your father, my dear Louisa, to know that you do not come to the consideration of that question with the previous habits of mind, and habits of life, that belong to many young women.'

'No, father,' she returned, 'I do not.'

'I now leave you to judge for yourself,' said Mr Gradgrind. 'I have stated the case, as such cases are usually stated among practical minds; I have stated it, as the case of your mother and myself was stated in its time. The rest, my dear Louisa, is for you to decide.'

From the beginning, she had sat looking at him fixedly. As he now leaned back in his chair, and bent his deep-set eyes upon her in his turn, perhaps he might have seen one wavering moment in her, when she was impelled to throw herself upon his breast, and give him the pent-up confidences of her heart. But, to see it, he must have overleaped at a bound the artificial barriers he had for many years been erecting, between himself and all those subtle essences of humanity which will elude the utmost cunning of algebra until the last trumpet ever to be sounded shall blow even algebra to wreck. The barriers were too many and too high for such a leap. With his unbending, utilitarian, matter-of-fact face, he hardened her again; and the moment shot away into the plumbless depths of the past, to mingle with all the lost opportunities that are drowned there.

Removing her eyes from him, she sat so long looking silently towards the town, that he said, at length: 'Are you consulting the chimneys of the Coketown works, Louisa?'

'There seems to be nothing there but languid and monotonous smoke. Yet when the night comes, Fire bursts out, father!' she answered, turning quickly.

'Of course I know that, Louisa. I do not see the application of the remark.' To do him justice he did not, at all.

She passed it away with a slight motion of her hand, and concentrating her attention upon him again, said, 'Father, I have often thought that life is very short.'—This was so distinctly one of his subjects that he interposed.

'It is short, no doubt, my dear. Still, the average duration of human life is proved to have increased of late years. The calculations of various life assurance and annuity offices, among other figures which cannot go wrong, have established the fact.'

'I speak of my own life, father.'

'O indeed? Still,' said Mr Gradgrind. 'I need not point out to you, Louisa, that it is governed by the laws which govern lives in the aggregate.'

'While it lasts, I would wish to do the little I can, and the little I am fit for. What does it matter?'

Mr Gradgrind seemed rather at a loss to understand the last four words; replying, 'How, matter? What matter, my dear?'

'Mr Bounderby,' she went on in a steady, straight way, without regarding this, 'asks me to marry him. The question I have to ask myself is, shall I marry him? That is so, father, is it not? You have told me so, father. Have you not?'

'Certainly, my dear.'

'Let it be so. Since Mr Bounderby likes to take me thus, I am satisfied to accept his proposal. Tell him, father, as soon as you please, that this was my answer. Repeat it, word for word, if you can, because I should wish him to know what I said.'

'It is quite right, my dear,' retorted her father approvingly, 'to be exact. I will observe your very proper request. Have you any wish in reference to the period of your marriage, my child?'

'None, father. What does it matter?'

Louisa, brought up by her 'eminently practical' father, accepts Bounderby's proposal on the rational and eminently practical grounds that the marriage will help her much-loved, ne'er-do-well brother. But her loveless marriage is doomed, and only a few years later 'the fire bursts out' when she meets a debonair young politician. She flees from him and from Mr Bounderby back to her father:

'How could you give me life, and take from me all the inappreciable things that raise it from the state of conscious death? Where are the graces of my soul? Where are the sentiments of my heart? What have you done, O father, what have you done, with the garden that should have bloomed once, in this great wilderness here?'

CHARLES DICKENS, *Hard Times*, 1854

He entered. He stood before me. What his words were you can imagine; his manner you can hardly realise, nor can I forget it. He made me, for the first time, feel what it costs a man to declare affection when he doubts response. . . .

CHARLOTTE BRONTË, in a letter

Before we moved to Lahore, Daddyji had gone to Mussoorie, a hill station in the United Provinces, without telling us why he was going out of the Punjab. Now, several months after he made that trip, he gathered us around him in the drawing room at 11 Temple Road while Mamaji mysteriously hurried Sister Pom upstairs. . . . He said that by right and tradition the oldest daughter had to be given in marriage first, and that the ripe age for marriage was nineteen. He said that when a girl approached that age her parents, who had to take the initiative, made many inquiries and followed many leads . . . 'That's why I said nothing to you children about why I went to Mussoorie,' he concluded. 'I went to see a young man for Pom. She's already nineteen.'

We were stunned. We have never really faced the idea that Sister Pom might get married and suddenly leave, I thought.

'We won't lose Pom, we'll get a new family member,' Daddyji said, as if reading my thoughts.

*

'Your mother has just taken Pom up to tell her,' Daddyji said. 'But she's a good girl. She will agree.' He added, 'The young man in question is twenty-eight years old. He's a dentist, and so has a profession.'

'Did you get a dentist because Sister Pom has bad teeth?' Usha asked. . . . Daddyji laughed. 'I confess I didn't think of anyone's teeth when I chose the young man in question.'

'What is he like?' I asked. 'What are we to call him?'

'He's a little bit on the short side, but he has a happy-go-lucky nature, like Nimi's. He doesn't drink, but, unfortunately, he does smoke. His father died at an early age of a heart attack, but he has a nice mother, who will not give Pom any trouble. It seems that everyone calls him Kakaji.'

We all laughed. Kakaji, or 'youngster,' was what very small boys were called. . . . In spite of myself, I pictured a boy smaller than I was and imagined him taking Sister Pom away, and then I imagined her having to keep his pocket money, to arrange his clothes in the cupboards, to comb his hair. My mouth felt dry.

'What will Kakaji call Sister Pom?' I asked.

'Pom, silly—what else?' Sister Umi said.

Mamaji and Sister Pom walked into the room. Daddyji made a place for Sister Pom next to him and said, 'Now, now, now, no reason to cry. Is it to be yes?'

'Whatever you say,' Sister Pom said in a small voice, between sobs.

'Pom, how can you say that? You've never seen him,' Sister Umi said.

*

'You promised me you wouldn't cry again,' Mamaji said to Sister Pom, patting her on the back, and then, to Daddyji, 'She's agreed.'

Daddyji said much else, sometimes talking just for the sake of talking, sometimes laughing at us because we were sniffling, and all the time trying to make us believe that this was a happy occasion. First, Sister Umi took issue with him: parents had no business arranging marriages; if she were Pom she would run away. Then Sister Nimi: all her life she had heard him say to us children, 'Think for yourself—be independent,' and here he was not allowing Pom to think for herself. Brother Om took Daddyji's part: girls who didn't get married became a burden on their parents, and Daddyji had four daughters to marry off, and would be retiring in a few years. Sisters Nimi and Umi retorted: they hadn't gone to college to get married off, to have some young man following them around like a leech. Daddyji just laughed. . . . 'Go and bless your big sister,' Mamaji said, pushing me in the direction of Sister Pom.

'I don't want to,' I said. 'I don't know him.'

'What'll happen to Sister Pom's room?' Usha asked. She and Ashok didn't have rooms of their own. They slept in Mamaji's room.

'Pom's room will remain empty, so that any time she likes she can come and stay in her room with Kakaji,' Daddyji said.

The thought that a man I never met would sleep in Pom's room with Sister Pom there made my heart race. A sob shook me. I ran outside.

*

Kakaji had left without formally committing himself. Then, four days later, when we were all sitting in the drawing room, a servant brought a letter to Mamaji. She told us that it was from Kakaji's mother, and that it asked if Sister Pom might be engaged to Kakaji. 'She even wants to know if Pom can be married in April or May,' Mamaji said excitedly.

*

'You still have time to change your mind,' Daddyji said to Sister Pom. 'What do you really think of him?'

Sister Pom wouldn't say anything.

'How do you expect her to know what her mind is when all that the two talked about was a picture and her bachelor's exam in May?' Sister Umi demanded. 'Could she have fallen in love already?'

'Love, Umi, means something very different from "falling in love,"' Daddyji said. 'It's not an act but a lifelong process. The best we can do as Pom's parents is to give her love every opportunity to grow.'

'But doesn't your "every opportunity" include knowing the person better than over a cup of tea, or whatever?' Sister Umi persisted.

'Yes, of course it does. But what we are discussing here is a simple matter of choice—not love,' Daddyji said. 'To know a person, to love a person, takes years of living together.'

'Do you mean, then, that knowing a person and loving a person are the same thing?' Sister Umi asked.

'Not quite, but understanding and respect are essential to love, and that cannot come from talking together, even over a period of days or months. That can come only in good time, through years of experience. It is only when Pom and Kakaji learn to consider each other's problems as one and the same that they will find love.'

VED MEHTA, *The Ledge Between the Streams*, 1982

'I came up,' he said, speaking curiously matter-of-fact and level, 'to ask if you'd marry me. You are free, aren't you?'

There was a long silence, whilst his blue eyes, strangely impersonal, looked into her eyes to seek an answer to the truth. He was looking for the truth out of her. And she, as if hypnotized, must answer at length.

'Yes, I am free to marry.'

The expression of his eyes changed, became less impersonal, as if he were looking almost at her, for the truth of her. Steady and intent and eternal they were, as if they would never change. They seemed to fix and to resolve her. She quivered, feeling herself created, will-less, lapsing into him, into a common will with him.

'You want me?' she said.

A pallor came over his face.

'Yes,' he said.

Still there was suspense and silence.

'No,' she said, not of herself. 'No, I don't know.'

He felt the tension breaking up in him, his fists slackened, he was unable to move. He stood looking at her, helpless in his vague collapse. For the moment she had become unreal to him. Then he saw her come to him, curiously direct and as if without movement, in a sudden flow. She put her hand to his coat.

'Yes I want to,' she said impersonally, looking at him with wide, candid, newly-opened eyes, opened now with supreme truth. He went very white as he stood, and did not move, only his eyes were held by hers, and he suffered. She seemed to see him with her newly-opened, wide eyes, almost of a child, and with a strange movement, that was agony to him, she reached slowly forward her dark face and her breast to him, with a slow insinuation of a kiss that made something break in his brain, and it was darkness over him for a few moments.

He had her in his arms, and, obliterated, was kissing her. And it was sheer, blenched agony to him, to break away from himself. She was there so small and light and accepting in his arms, like a child, and yet with such an insinuation of embrace, of infinite embrace, that he could not bear it, he could not stand.

He turned and looked for a chair, and keeping her still in his arms, sat down with her close to him, to his breast. Then, for a few seconds, he went utterly to sleep, asleep and sealed in the darkest sleep, utter, extreme oblivion.

D. H. LAWRENCE, *The Rainbow*, 1915

I go about murmuring, 'I have made that dignified girl *commit* herself, I have, I have', and then I vault over the sofa with exultation.

WALTER BAGEHOT, from a letter to Elizabeth Wilson, 1857

¶ (*They were married the following year.*)

ALGERNON: I hope, Cecily, I shall not offend you if I state quite frankly and openly that you seem to me to be in every way the visible personification of absolute perfection.

CECILY: I think your frankness does you great credit, Ernest. If you will allow me, I will copy your remarks into my diary. (*Goes over to table and begins writing in diary.*)

ALGERNON: Do you really keep a diary? I'd give anything to look at it. May I?

CECILY: Oh no. (*Puts her hand over it.*) You see, it is simply a very young girl's record of her own thoughts and impressions, and consequently meant for publication. When it appears in volume form I hope you will order a copy. But pray, Ernest, don't stop. I delight in taking down from dictation. I have reached 'absolute perfection'. You can go on. I am quite ready for more.

ALGERNON (*somewhat taken aback*): Ahem! Ahem!

CECILY: Oh, don't cough, Ernest. When one is dictating one should speak fluently and not cough. Besides, I don't know how to spell a cough. (*Writes as* ALGERNON *speaks.*)

ALGERNON (*speaking very rapidly*): Cecily, ever since I first looked upon your wonderful and incomparable beauty, I have dared to love you wildly, passionately, devotedly, hopelessly.

CECILY: I don't think that you should tell me that you love me wildly, passionately, devotedly, hopelessly. Hopelessly doesn't seem to make much sense, does it?

ALGERNON: Cecily!

[*Enter* MERRIMAN.

MERRIMAN: The dog-cart is waiting, sir.

ALGERNON: Tell it to come round next week, at the same hour.

MERRIMAN (*looks at* CECILY, *who makes no sign*): Yes, sir.

[MERRIMAN *retires.*

CECILY: Uncle Jack would be very much annoyed if he knew you were staying on till next week, at the same hour.

ALGERNON: Oh, I don't care about Jack. I don't care for anybody in the whole world but you. I love you, Cecily. You will marry me, won't you?

CECILY: You silly boy! Of course. Why, we have been engaged for the last three months.

ALGERNON: For the last three months?

CECILY: Yes, it will be exactly three months on Thursday.

ALGERNON: But how did we become engaged?

CECILY: Well, ever since dear Uncle Jack first confessed to us that he had a younger brother who was very wicked and bad, you, of course, have formed the chief topic of conversation between myself and Miss Prism. And, of course, a man who is much talked about is always very attractive. One feels there must be something in him, after all. I dare say it was foolish of me, but I fell in love with you, Ernest.

ALGERNON: Darling. And when was the engagement actually settled?

CECILY: On the 14th of February last. Worn out by your entire ignorance of my existence, I determined to end the matter one way or the other, and after a long struggle with myself I accepted you under this dear old tree here. The next day I bought this little ring in your name, and this is the little bangle with the true lover's knot I promised you always to wear.

ALGERNON: Did I give you this? It's very pretty, isn't it?

CECILY: Yes, you've wonderfully good taste, Ernest. It's the excuse

I've always given for your leading such a bad life. And this is the box in which I keep all your dear letters. (*Kneels at table, opens box, and produces letters tied up with blue ribbon.*)

ALGERNON: My letters! But, my own sweet Cecily, I have never written you any letters.

CECILY: You need hardly remind me of that, Ernest. I remember only too well that I was forced to write your letters for you. I wrote always three times a week, and sometimes oftener.

ALGERNON: Oh, do let me read them, Cecily?

CECILY: Oh, I couldn't possibly. They would make you far too conceited. (*Replaces box.*) The three you wrote me after I had broken off the engagement are so beautiful, and so badly spelled, that even now I can hardly read them without crying a little.

ALGERNON: But was our engagement ever broken off?

CECILY: Of course it was. On the 22nd of last March. You can see the entry if you like. (*Shows diary.*) 'To-day I broke off my engagement with Ernest. I feel it is better to do so. The weather still continues charming.'

ALGERNON: But why on earth did you break it off? What had I done? I had done nothing at all. Cecily, I am very much hurt indeed to hear you broke it off. Particularly when the weather was so charming.

CECILY: It would hardly have been a really serious engagement if it hadn't been broken off at least once. But I forgave you before the week was out.

ALGERNON (*crossing to her, and kneeling*): What a perfect angel you are, Cecily.

CECILY: You dear romantic boy. (*He kisses her, she puts her fingers through his hair.*) I hope your hair curls naturally, does it?

ALGERNON: Yes, darling, with a little help from others.

CECILY: I am so glad.

ALGERNON: You'll never break off our engagement again, Cecily?

CECILY: I don't think I could break it off now that I have actually met you. Besides, of course, there is the question of your name.

ALGERNON: Yes, of course. (*Nervously.*)

CECILY: You must not laugh at me, darling, but it had always been a girlish dream of mine to love some one whose name was Ernest.

[ALGERNON *rises,* CECILY *also.*

There is something in that name that seems to inspire absolute confidence. I pity any poor married woman whose husband is not called Ernest.

ALGERNON: But, my dear child, do you mean to say you could not love me if I had some other name?

CECILY: But what name?

ALGERNON: Oh, any name you like—Algernon—for instance . . .

CECILY: But I don't like the name of Algernon.

ALGERNON: Well, my own dear, sweet, loving little darling, I really can't see why you should object to the name of Algernon. It is not at all a bad name. In fact, it is rather an aristocratic name. Half of the chaps who get into the Bankruptcy Court are called Algernon. But seriously, Cecily—(*moving to her*)—if my name was Algy, couldn't you love me?

CECILY (*rising*): I might respect you, Ernest, I might admire your character, but I fear that I should not be able to give you my undivided attention.

ALGERNON: Ahem! Cecily! (*Picking up hat.*) Your Rector here is, I suppose, thoroughly experienced in the practice of all the rites and ceremonials of the Church?

CECILY: Oh, yes. Dr Chasuble is a most learned man. He has never written a single book, so you can imagine how much he knows.

ALGERNON: I must see him at once on a most important christening—I mean on most important business.

CECILY: Oh!

ALGERNON: I shan't be away more than half an hour.

CECILY: Considering that we have been engaged since February the 14th, and that I only met you to-day for the first time, I think it is rather hard that you should leave me for so long a period as half an hour. Couldn't you make it twenty minutes?

ALGERNON: I'll be back in no time. (*Kisses her and rushes down the garden.*)

CECILY: What an impetuous boy he is! I like his hair so much. I must enter his proposal in my diary.

OSCAR WILDE, *The Importance of Being Earnest*, 1895

A PROPOSALE

Next morning while imbibing his morning tea beneath his pink silken quilt Bernard decided he must marry Ethel with no more delay. I love the girl he said to himself and she must be mine but I somehow feel I can not propose in London it would not be seemly in the city of London. We must go for a day in the country and when surrounded by the gay twittering of the birds and the smell of the cows I will lay my suit at her feet and he waved his arm wildly at the gay thought. Then he sprang from bed and gave a rat tat at Ethel's door.

Are you up my dear he called.

Well not quite said Ethel hastilly jumping from her downy nest.

Be quick cried Bernard I have a plan to spend a day near Windsor Castle and we will take our lunch and spend a happy day.

Oh Hurrah shouted Ethel I shall soon be ready as I had my bath last night so wont wash very much now.

No dont said Bernard and added in a rarther fervent tone through the chink of the door you are fresher than the rose my dear no soap could make you fairer.

Then he dashed off very embarrased to dress. Ethel blushed and felt a bit excited as she heard the words and she put on a new white muslin dress in a fit of high spirits. She looked very beautifull with some red roses in her hat and the dainty red ruge in her cheeks looked quite the thing. Bernard heaved a sigh and his eyes flashed as he beheld her and Ethel thorght to herself what a fine type of manhood he reprisented with his nice thin legs in pale broun trousers and well fitting spats and a red rose in his button hole and rarther a sporting cap which gave him a great air with its quaint check and little flaps to pull down if necessary. Off they started the envy of all the waiters.

They arrived at Windsor very hot from the jorney and Bernard at once hired a boat to row his beloved up the river. Ethel could not row but she much enjoyed seeing the tough sunburnt arms of Bernard tugging at the oars as she lay among the rich cushons of the dainty boat. She had a rarther lazy nature but Bernard did not know of this. However he soon got dog tired and sugested lunch by the mossy bank.

Oh yes said Ethel quickly opening the sparkling champaigne.

Dont spill any cried Bernard as he carved some chicken.

They eat and drank deeply of the charming viands ending up with merangs and choclates.

Let us now bask under the spreading trees said Bernard in a passiunate tone.

Oh yes lets said Ethel and she opened her dainty parasole and sank down upon the long green grass. She closed her eyes but she was far from asleep. Bernard sat beside her in profound silence gazing at her pink face and long wavy eye lashes. He puffed at his pipe for some moments while the larks gaily caroled in the blue sky. Then he edged a trifle closer to Ethels form.

Ethel he murmered in a trembly voice.

Oh what is it said Ethel hastily sitting up.

Words fail me ejaculated Bernard horsly my passion for you is intense he added fervently. It has grown day and night since I first beheld you.

Oh said Ethel in supprise I am not prepared for this and she lent back against the trunk of the tree.

Bernard placed one arm tightly round her. When will you marry me Ethel he uttered you must be my wife it has come to that I love you so intensly that if you say no I shall perforce dash my body to the brink of yon muddy river he panted wildly.

Oh dont do that implored Ethel breathing rarther hard.

Then say you love me he cried.

Oh Bernard she sighed fervently I certinly love you madly you are to me like a Heathen god she cried looking at his manly form and handsome flashing face I will indeed marry you.

How soon gasped Bernard gazing at her intensly.

As soon as possible said Ethel gently closing her eyes.

My Darling whispered Bernard and he seiezed her in his arms we will be marrid next week.

Oh Bernard muttered Ethel this is so sudden.

No no cried Bernard and taking the bull by both horns he kissed her violently on her dainty face. My bride to be he murmered several times.

Ethel trembled with joy as she heard the mistick words.

Oh Bernard she said little did I ever dream of such as this and she suddenly fainted into his out stretched arms.

Oh I say gasped Bernard and laying the dainty burden on the grass he dashed to the waters edge and got a cup full of the fragrant river to pour on his true loves pallid brow.

She soon came to and looked up with a sickly smile Take me back to the Gaierty hotel she whispered faintly.

With plesure my darling said Bernard I will just pack up our viands ere I unloose the boat.

Ethel felt better after a few drops of champagne and began to tidy her hair while Bernard packed the remains of the food. Then arm in arm they tottered to the boat.

I trust you have not got an illness my darling murmered Bernard as he helped her in.

Oh no I am very strong said Ethel I fainted from joy she added to explain matters.

Oh I see said Bernard handing her a cushon well some people do he added kindly and so saying they rowed down the dark stream now flowing silently beneath a golden moon. All was silent as the lovers glided home with joy in their hearts and radiunce on their faces only the sound of the mystearious water lapping against the frail vessel broke the monotony of the night.

DAISY ASHFORD, *The Young Visiters*, 1919.

¶ *Daisy Ashford wrote this when she was a small girl, and discovered the manuscript years later, in 1919, when it was published with the original spelling and punctuation intact.*

I was in rare fettle and the heart had touched a new high. I don't know anything that braces one up like finding you haven't got to get married after all.

P. G. WODEHOUSE, *Jeeves in the Offing*, 1960

the day I got him to propose to me yes first I gave him the bit of seedcake out of my mouth and it was leapyear like now yes 16 years ago my God after that long kiss I near lost my breath yes he said I was a flower of the mountain yes so we are flowers all a womans body yes that was one true thing he said in his life and the sun shines for you today yes that was why I liked him because I saw he understood or felt what a woman is and I knew I could always get round him and I gave him all the pleasure I could leading him on till he asked me to say yes and I wouldnt answer first only looked out over the sea and the sky I was thinking of so many things he didnt know of Mulvey and Mr Stanhope and Hester and father and old captain Groves and the sailors playing all birds fly and I say stoop and washing up dishes they called it on the pier and the sentry in front of the governors house with the thing round his white helmet poor devil half roasted and the Spanish girls laughing in their shawls and their tall combs and the auctions in the morning the Greeks and the jews and the Arabs and the devil knows who else from all the ends of Europe and Duke street and the fowl market all clucking outside Larby Sharons and the poor donkeys slipping half asleep and the vague fellows in the cloaks asleep in the shade on the steps and the big wheels of the carts of the bulls and the old castle thousands of years old yes and those handsome Moors all in white and turbans like kings asking you to sit down in their little bit of a shop and Ronda with the old windows of the posadas glancing eyes a lattice hid for her lover to kiss the iron and the wineshops half open at night and the castanets and the night we missed the boat at Algeciras the watchman going about serene with his lamp and O that awful deep-down torrent O and the sea the sea crimson sometimes like fire and the glorious sunsets and the figtrees in the Alameda gardens yes and all the queer little streets and pink and blue and yellow houses and the rosegardens and the jessamine and geraniums and cactuses and Gibraltar as a girl where I was a Flower of the mountain yes when I put the rose in my hair like the Andalusian girls used or shall I wear a red yes and how he kissed me under the Moorish wall and I thought well as well him as another and then I asked him with my eyes to ask again yes and then he asked me would I yes to say yes my mountain flower and first I put my arms around him yes and drew him down to me so he could feel my breasts all perfume yes and his heart was going like mad and yes I said yes I will Yes.

JAMES JOYCE, *Ulysses*, 1922

3

With this Ring I thee Wed

THE wedding service from the 1549 Church of England Prayer Book remains the most resonant exchange of marriage vows in the English language. There have been many attempts to devise contemporary or lay versions, but none has the beauty of the original.

The woman's promise to 'obey' now seems absurdly outdated, but there is nothing new in the modern practice of omitting it. In 1832 Robert Owen, son of the Utopian humanist, made this fierce declaration at his wedding:

Of the unjust rights which in virtue of this ceremony an iniquitous law tacitly gives me over the person and property of another, I cannot legally, but I can morally, divest myself. And I hereby distinctly and emphatically declare that I consider myself, and earnestly desire to be considered by others, as utterly divested, now and during the rest of my life, of any such rights, the barbarous relics of a feudal, despotic system.

And Lucy Stone, an early American feminist, declared jointly with her husband at their wedding in 1855:

While we acknowledge our mutual affection by publicly assuming the relationship of husband and wife . . . we deem it a duty to declare that this act on our part implies no sanction of, nor promise of voluntary obedience to such of the present laws of marriage as refuse to recognize the wife as an independent, rational being, while they confer upon the husband an injurious and unnatural superiority . . .

(from Eleanor Flexner, *Centuries of Struggle*).

The long tradition of epithalamiums celebrates weddings as moments for rejoicing. But there are also occasions when one of the couple, or some of the bystanders, view the marriage with justified misgiving. 'He was inclined to inquire what he had done, or she lost, for that matter, that he deserved to be caught in a gin which would cripple him, if not her also, for the rest of a lifetime' (Thomas Hardy, *Jude the Obscure*).

Even if the day is joyful for the couple, the 'extras' often experience mixed feelings, ranging from the gentle nostalgia of Emma Bovary's father to the painful memories evoked for Queen Victoria by her son's wedding, to the wrenching sense of loss and separation suffered by Dorothy Wordsworth while her beloved brother William was being married. A wedding is always a moment of dynamic change, not just for the couple, but for the entire constellation of their family and friends, 'with all the power that being changed can give'.

THE FORM OF SOLEMNIZATION OF MATRIMONY

Wilt thou have this Woman to thy wedded wife, to live together after God's ordinance in the holy estate of Matrimony? Wilt thou love her, comfort her, honour, and keep her in sickness and in health: and, forsaking all other, keep thee only unto her, so long as ye both shall live?

The Man shall answer,

I will.

Then shall the Priest say unto the woman,

Wilt thou have this Man to thy wedded husband, to live together after God's ordinance in the holy estate of Matrimony? Wilt thou obey him, and serve him, love, honour, and keep him in sickness and in health: and, forsaking all other, keep thee only unto him, so long as ye both shall live?

The Woman shall answer,

I will.

*

I [name] take thee [name] to my wedded wife, to have and to hold from this day forward, for better for worse, for richer for poorer, in sickness and in health, to love and to cherish, till death us do part, according to God's holy ordinance: and thereto I plight thee my troth.

Then shall they loose their hands; and the Woman, with her right hand taking the Man by his right hand, shall likewise say after the Minister,

I [name] take thee [name] to my wedded husband, to have and to hold from this day forward, for better for worse, for richer for poorer, in sickness and in health, to love, cherish, and to obey, till death us do part, according to God's holy ordinance: and thereto I give thee my troth.

*

With this Ring I thee wed, with my body I thee worship, and with all my worldly goods I thee endow: In the Name of the Father, and of the Son, and of the Holy Ghost. Amen.

The Book of Common Prayer, 1552

Guardian of Helicon, Urania's son,
Come down: carry the gentle bride
Off to her husband. *O Hymen,*
Hymenaeus, come!
Garland your hair with marjoram,
Soft-scented; veil your face and come
Smiling down to us, saffron shoes
On milk-white feet.
Awakened on this happy day,
Join us in lusty marriage-songs,
Join us in dancing, holding high
The marriage-torch.

*

Bridesmaids-in-waiting, gather now
And lift harmonious voices
In the wedding-song: *O Hymen,*
Hymen, Hymenaeus O!
Call his name, and summon him down
To bless this wedding-day: Hymen
Forerunner of Venus, lord
Of the wedding-night.
You are the only lord of lovers:
Which other god can men prefer
Before you, lord? *O Hymen,*
Hymen, Hymenaeus O!
The nervous father waits for you,
The bride, laying her girlhood down,
The eager groom, his ears alert
For your footstep.
You will pluck the bride like a flower
From her mother's arms, and lay her
In his proud young arms. *O Hymen,*
Hymen, Hymenaeus O!

Without your aid, there is no love
To leaven the wedding ritual
And give increase. Which of the gods
 Can equal you?
Without your aid, the house is bare
Of children; parents are barren
Without your aid. Which of the gods
 Can equal you?
Without your aid, the land lies bare
Of sons to guard it: only you
Can give it strength. Which of the gods
 Can equal you?

Open the doors; the bride is here.
See how the wedding-torches shake
Their heads, their fiery locks flying free
 To welcome her?
A shy young bride, half bashful
Half excited; she hesitates
Until love's calling draws her on
 Melting in tears.
No tears today! No girl on earth
So beautiful; no girl will see
A fairer wedding-day than this
 Dawn in the east.
Like you, the hyacinth, apart
From every flower in a rich man's
Pleasure garden: no more delay,
 The day is waiting.
Come forward, sweet, and hear our song;
Our torches dip their golden heads
To welcome you. This day is yours—
 Come forward, sweet.
No other woman's bed will lure
Your husband, no furtive lusts will draw
Him away from his soft repose
 Between your breasts,
But like the tree trunk gathered close
By the soft embracing vine,
So he in you. This day is yours—
 Come forward, sweet.

The marriage-couch, its polished legs,
Your lord's delight at hot mid-day
Or in the fleeting night; this day
    Is yours—come, sweet.
Lift high the torches: let them blaze
Bright as the wedding-veil. Come, sing
The wedding song. *O Hymen,*
    Hymen, Hymenaeus O!

*

With golden steps, lady, beloved
Of the gods, step forward, cross
The polished threshold. *O Hymen,*
    Hymen, Hymenaeus O!
Your lord waits on a royal couch
(Peep in and see): waits, rising
To welcome you. *O Hymen,*
    Hymen, Hymenaeus O!
His heart is on fire with love,
As yours is: his passion burns
Inside, deep down. *O Hymen,*
    Hymen, Hymenaeus O!
Page-boy, let go her arm, her soft
Bride's arm; now she must go alone
To her husband's bed. *O Hymen,*
    Hymen, Hymenaeus O!
Ladies-of-honour, worthy wives
Of worthy husbands, old in love,
Escort her in. *O Hymen,*
    Hymen, Hymenaeus O!
Come now, husband; your wife waits
Pale as a white convolvulus
In the marriage-bed. *O Hymen,*
    Hymen, Hymenaeus O!
Come, handsome youth, as blessed as she
With loveliness: come forward now.
Venus is with you: forward now,
    The day is dying.
Come in, come in! Your love is there
For all to see. Venus is with you,

Desire in you made manifest.
    No hiding now.
Far easier to count each grain of sand
In Africa, or tally the bright stars,
Than reckon up the games of love
    You two will play.
Play on! Love's games make children,
Children an ancient line as proud
As yours demands. Play on, and sow
    The future's seed.
Soon may a baby Manlius
Reach out soft hands from his mother's breast,
Laugh at his father, baby lips
    Parted in joy.
May he grow so like his father
That all strangers acknowledge him
Noble as Manlius, modest
    As Manlius' wife,
The son honoured in his mother
As one other only, the son
Of Penelope flower of wives,
    Telemachus.

Girls, close the doors now. Our part
Is over. And you, happy pair,
God bless you. Practise your nimble youth
    In the gods' undying gift.

CATULLUS, *c.*60 BC

EPITHALAMIUM

This girl all in white is my crystal of light
Kissed by heaven to earth in a dancing gift
Of a bride in her freshness, whom youth and love lift,
With two sunbeams for bridesmaids, their father's delight.

I have married my bride in a ring of green fields
Round a church on a hill where all nature's her dress,
While below the bright lake reflects heaven's caress
With a leap of the organ and towered bells' peals.

The sheep stray up close leaving wool on the hedges,
Their gifts for her future; town friends are our guests.
Here the wheat, there the barley wave from the lanes' edges,
And housemartins flash bringing song from their nests.

The rice will be scattered, her blossom will glow
As new garland and sheaf and the berry-bright trees,
And the laughter of friends, and the families grow;
And our spring-wind will dance on the grandparents' knees.

They say that on the bridal-day
(Unless I am mistaken)
The music of the spheres is heard
As holiest vows are taken
(And I can hear them now. Can you?)
And all the plants awaken
And stay awake the whole night through
Till morning sheets are shaken.

The pink-turned-blue forget-me-not
With no stalks on its leaves;
The sun-gold lady's bedstraw
(No sting nor prickle leaves);
Love-in-a-mist sky-blue and white,
Soft maiden-hair and comfrey—
These will delight my girl tonight
With heartsease in the country.

Later today we'll be scattered away,
But something is altered for good:
Here death is defeated by life-giving love,
And light conquers dark, as it should.
May our joy and our thanks race the blood in your flanks
For, from where you have come, far and wide,
You have honoured and blest of all moments the best:
When a man takes his girl as his bride.

FRANCIS WARNER, 1983

A BALLADE UPON A WEDDING

I tell thee *Dick* where I have been,
Where I the rarest things have seen;
  Oh things without compare!
Such sights cannot be found
In any place on English ground
  Be it at Wake, or Fair.

At *Charing-Crosse*, hard by the way
Where we (thou know'st) do sell our Hay,
  There is a house with stairs;
And there did I see comming down
Such folks as are not in our Town,
  Vorty at least, in Pairs.

Amongst the rest, one Pest'lent fine,
(His beard no bigger though then thine)
  Walkt on before the rest:
Our Landlord looks like nothing to him:
The King (God blesse him) 'twould undo him
  Should he go still so drest.

At Course-a-Park, without all doubt,
He should have first been taken out
  By all the maids i'th Town:
Though lusty *Roger* there had been,
Or little *George* upon the Green,
  Or *Vincent* of the Crown.

But wot you what? the youth was going
To make an end of all his woing;
  The Parson for him staid:
Yet by his leave (for all his haste)
He did not so much wish all past
  (Perchance) as did the maid.

The maid (and thereby hangs a tale)
For such a maid no Whitson-ale
  Could ever yet produce:

No Grape that's kindly ripe, could be
So round, so plump, so soft as she,
  Nor half so full of Juyce.

Her finger was so small, the Ring
Would not stay on which they did bring,
  It was too wide a Peck:
And to say truth (for out it must)
It lookt like the great Collar (just)
  About our young Colt's neck.

Her feet beneath her Petticoat,
Like little mice stole in and out,
  As if they fear'd the light:
But oh! she dances such a way!
No sun upon an Easter day
  Is half so fine a sight.

He would have kist her once or twice,
But she would not, she was so nice,
  She would not do't in sight,
And then she lookt as who should say
I will do what I list today;
  And you shall do't at night.

Her Cheeks so rare a white was on,
No Dazy makes comparison,
  (Who sees them is undone)
For streaks of red were mingled there,
Such as on a Katherne Pear,
  (The side that's next the Sun.)

Her lips were red, and one was thin,
Compar'd to that was next her Chin;
  (Some Bee had stung it newly.)
But (*Dick*) her eyes so guard her face;
I durst no more upon them gaze
  Then on the sun in *July*.

Her mouth so small when she does speak,
Thou'dst swear her teeth her words did break
  That they might passage get,
But she so handled still the matter,
They came as good as ours, or better,
  And are not spent a whit.

If wishing should be any sin,
The Parson himself had guilty bin;
  (she lookt that day so purely,)
And did the Youth so oft the feat
At night, as some did in conceit,
  It would have spoiled him, surely.

*

Now hatts fly off, and youths carouse;
Healths first go round, and then the house,
  The Brides came thick and thick:
And when 'twas nam'd another health,
Perhaps he made it hers by stealth,
  (and who could help it? *Dick*)

O'th'sodain up they rise and dance;
Then sit again and sigh, and glance:
  Then dance again and kisse:
Thus sev'ral waies the time did passe,
Till ev'ry Woman wisht her place,
  And ev'ry Man wisht his.

By this time all were stoln aside
To counsel and undresse the Bride;
  But that he must not know:
But yet 'twas thought he ghest her mind
And did not mean to stay behind
  Above an hour or so.

When in he came (*Dick*) there she lay
Like new-faln snow melting away,
  ('Twas time I trow to part)
Kisses were now the onely stay,

Which soon she gave, as who would say,
Good Boy! with all my heart.

But just as leav'ns would have to crosse it,
In came the Bridesmaids with the Posset:
The Bridegroom eat in spight;
For had he left the Women to't
It would have cost two hours to do't,
Which were too much that night.

At length the candles out and out,
All that they had not done, they do't;
What that is, who can tell?
But I beleeve it was no more
Then thou and I have done before
With *Bridget*, and with *Nell*.

SIR JOHN SUCKLING, 1646

The guests arrived early, in carriages, one-horse traps, two-wheeled wagonettes, old cabs minus their hoods, or spring-vans with leather curtains. The young people from the neighbouring villages came in farm-carts, standing up in rows, with their hands on the rails to prevent themselves from falling: trotting along and getting severely shaken up. Some people came from thirty miles away, from Goderville, Normanville and Cany. All the relatives on both sides had been invited. Old estrangements had been patched up and letters sent to long-forgotten acquaintances.

Every now and then the crack of a whip would be heard behind the hedge. Another moment, and the gate opened, a trap drove in. It galloped up to the foot of the steps, pulled up sharp and emptied its load. They tumbled out on all sides, with a rubbing of knees and a stretching of arms. The ladies were in bonnets, and wore town-style dresses, with gold watch-chains, and capes with the ends tucked inside their sashes, or little coloured neckerchiefs pinned down at the back, leaving their necks bare. The boys were dressed like their papas, and looked uncomfortable in their new clothes. Many, indeed, were that day sampling the first pair of boots they had ever had. Beside them went a big girl of fourteen to sixteen, cousin or elder sister no doubt, all red and flustered and speechless in her white first-communion dress (let down for the occasion), with a rose pomade smarmed over her hair, and gloves that she was terrified of getting dirty. As there were not enough stablemen to unharness all the carriages, the gentlemen pulled up their sleeves and went to work themselves. According to the difference in their social status, they wore dress-coats, frock-coats, jackets or waistcoats.

*

The mayor's office being within a mile and a half of the farm, they made their way there on foot, and returned in the same manner after the church ceremony was over. At first the procession was compact, a single band of colour billowing across the fields, all along the narrow path that wound through the green corn; but soon it lengthened out and split up into several groups, which dawdled to gossip. The fiddler led the way, his violin adorned with rosettes and streamers of ribbon. After him came the bride and bridegroom, then the relatives, then the friends, in any order; and the children kept at the back, amusing themselves by plucking the bell-flowers from among the oat-stalks, or playing among themselves without being seen. Emma's dress was too long and dragged on the ground slightly. Now and again she stopped to pull it up, and then with her gloved fingers she daintily removed the coarse grasses and thistle burrs, while Charles waited, empty-handed, until she had finished.

The wedding-feast had been laid in the cart-shed. On the table were four sirloins, six dishes of hashed chicken, some stewed veal, three legs of mutton, and in the middle a nice roast sucking-pig flanked by four pork sausages with sorrel. Flasks of brandy stood at the corners. A rich foam had frothed out round the corks of the cider-bottles. Every glass had already been filled to the brim with wine. Yellow custard stood in big dishes, shaking at the slightest jog of the table, with the initials of the newly wedded couple traced on its smooth surface in arabesques of sugared almond. For the tarts and confectioneries they had hired a pastry-cook from Yvetot. He was new to the district, and so had taken great pains with his work. At dessert he brought in with his own hands a tiered cake that made them all cry out. It started off at the base with a square of blue cardboard representing a temple with porticoes and colonnades, with stucco statuettes all round it in recesses studded with gilt-paper stars; on the second layer was a castle-keep in Savoy cake, surrounded by tiny fortifications in angelica, almonds, raisins and quarters of orange; and finally, on the uppermost platform, which was a green meadow with rocks, pools of jam and boats of nutshell, stood a little Cupid, poised on a chocolate swing whose uprights had two real rosebuds for knobs at the top.

The meal went on till dusk. When they got tired of sitting, they went for a stroll round the farm, or played a game of 'corks' in the granary, after which they returned to their seats. Towards the end some of the guests were asleep and snoring. But with the coffee there was a general revival. They struck up a song, performed party-tricks, lifted weights, went 'under your thumb', tried hoisting the carts up on their shoulders, joked broadly and kissed the ladies. The horses, gorged to the nostrils with oats, could hardly be got into the shafts to go home at night. They kicked and reared and broke their harness, their masters swore or laughed; and all through the night by the light of the moon there were runaway carriages galloping along country roads, plunging into ditches, careering over stone-heaps, bumping into the banks, while women leaned out of the doors trying to catch hold of the reins.

*

The bride had begged her father to be spared the customary pleasantries. However, a fishmonger cousin—the same who had brought a pair of soles for his wedding-present—was about to squirt some water out of his mouth through the keyhole of their room, when Roualt came up just in time to prevent him and explain that his son-in-law's position in the world forbade such improprieties. Only with reluctance did the cousin yield to that argument.

*

Charles had no facetious side, and hadn't shone at his wedding. He responded feebly to the puns and quips and innuendoes, the compliments and ribaldries, which it was considered necessary to let fly at him as soon as the soup appeared.

Next morning, however, he seemed a different man. He was the one you would have taken for yesterday's virgin. Whereas the bride gave nothing away. The slyest among them were nonplussed. They surveyed her as she approached with the liveliest curiosity. But Charles made no pretences. He called her 'my wife', addressed her affectionately, kept asking everyone where she was, looked for her everywhere, and frequently led her out into the yard, where he was seen, away among the trees, putting his arm round her waist, leaning over her as they walked, and burying his face in the frills of her bodice.

Two days after the wedding, husband and wife departed. Charles could not be away from his patients any longer. Roualt let them have his trap for the journey, and accompanied them himself as far as Vassonville. There he kissed his daughter for the last time, then got down and started back homewards. When he had gone about a hundred yards, he halted; and the sight of the trap going farther and farther into the distance, its wheels turning in the dust, drew from him a deep sigh. He remembered his own wedding, his early married life, his wife's first pregnancy. He had been pretty happy himself, that day he took her from her father and brought her home mounted behind him, trotting over the snow. For it had been round about Christmas-time, and the country was all white. She held on to him with one arm, her basket hung from the other; the wind lifted the long lace ribbons of her Caux headdress, so that sometimes they flapped over his mouth; and when he turned his head, he saw, close beside him against his shoulder, her rosy little face quietly smiling beneath the gold crown of her bonnet. . . . How long ago it was, all that. . . . He looked back, and there was nothing to be seen along the road. He felt dreary as an empty house. In his brain, still clouded with the fumes of the feast, dark thoughts and tender memories mingled. He felt a momentary inclination to go round by the church. But he was afraid the sight of it would only make him sadder still, and he went straight home.

GUSTAVE FLAUBERT, *Madame Bovary*, 1857

A note to Albert from Queen Victoria on the morning of their wedding:

10 February 1840

How are you to-day, and have you slept well? I have rested very well, and feel very comfortable to-day. What weather! I believe, however, the rain will cease.

Send one word when you, my most dearly loved bridegroom, will be ready. Thy ever-faithful, Victoria R.

JOURNAL 10 February 1840

Got up at a $\frac{1}{4}$ to 9—well, and having slept well; and breakfasted at $\frac{1}{2}$ p. 9. Mamma came before and brought me a Nosegay of orange flowers. My dearest kindest Lehzen gave me a dear little ring . . . Had my hair dressed and the wreath of orange flowers put on. Saw Albert for the last time alone, as my Bridegroom.

Saw Uncle, and Ernest whom dearest Albert brought up. At $\frac{1}{2}$ p. 12 I set off, dearest Albert having gone before. I wore a white satin gown with a very deep flounce of Honiton lace, imitation of old. I wore my Turkish diamond necklace and earrings, and Albert's beautiful sapphire brooch. . . . The Ceremony was very imposing, and fine and simple, and I think ought to make an everlasting impression on every one who promises at the Altar to keep what he or she promises. Dearest Albert repeated everything very distinctly. I felt so happy when the ring was put on, and by Albert. As soon as the Service was over, the Procession returned as it came, with the exception that my beloved Albert led me out. The applause was very great, in the Colour Court as we came through; Lord Melbourne, good man, was very much affected during the Ceremony and at the applause . . . I then returned to Buckingham Palace alone with Albert; they cheered us really most warmly and heartily; the crowd was immense; and the Hall at Buckingham Palace was full of people; they cheered us again and again . . . I went and sat on the sofa in my dressing-room with Albert; and we talked together there from 10 m. to 2 till 20 m. p. 2. Then we went downstairs where all the Company was assembled and went into the dining-room—dearest Albert leading me in . . . Talked to all after the breakfast, and to Lord Melbourne, whose fine coat I praised.

I went upstairs and undressed and put on a white silk gown trimmed with swansdown, and a bonnet with orange flowers. Albert went downstairs and undressed.

*

As soon as we arrived [at Windsor] we went to our rooms; my large dressing room is our sitting room; the 3 little blue rooms are his . . . After looking about our rooms for a little while, I went and changed my gown, and then came back to his small sitting room where dearest Albert was sitting and playing; he had put on his windsor coat; he took me on his knee, and kissed me and was so dear and kind. We had our dinner in our sitting room; but I had such a sick headache that I could eat nothing, and was obliged to lie down in the middle blue room for the remainder of the evening, on the sofa, but, ill or not, I never, never spent such an evening . . . He called me names of tenderness, I have never yet heard used to me before—was bliss beyond belief! Oh! this was the happiest day of my life!—May God help me to do my duty as I ought and be worthy of such blessings.

11 February 1840

When day dawned (for we did not sleep much) and I beheld that beautiful angelic face by my side, it was more than I can express! He does look so beautiful in his shirt only, with his beautiful throat seen. We got up at ¼ p. 8. When I had laced I went to dearest Albert's room, and we breakfasted together. He had a black velvet jacket on, without any neckcloth on, and looked more beautiful than it is possible for me to say . . . At 12 I walked out with my precious Angel, all alone—so delightful, on the Terrace and new Walk, arm in arm! . . . We talked a great deal together. We came home at one, and had luncheon soon after. Poor dear Albert felt sick and uncomfortable, and lay down in my room . . . He looked so dear, lying there and dozing.

To King Leopold 11 February 1840

I write to you from here [Windsor] the happiest, happiest Being that ever existed. Really, I do not think it possible for any one in the world to be happier, or as happy as I am. He is an Angel, and his kindness and affection for me is really touching. To look in those dear eyes, and that dear sunny face, is enough to make me adore him. What I can do to make him happy will be my greatest delight. Independent of my great personal happiness, the reception we both met with yesterday was the most gratifying and enthusiastic I ever experienced; there was no end of the crowds in London, and all along the road. I was a good deal tired last night.

Journal 12 February 1840

Already the 2nd day since our marriage; his love and gentleness is beyond everything, and to kiss that dear soft cheek, to press my lips to his, is heavenly bliss. I feel a purer more unearthly feel than I ever did. Oh! was ever woman so blessed as I am.

13 February 1840

My dearest Albert put on my stockings for me. I went in and saw him shave; a great delight for me.

Twenty-three years later, and two years after the death of her 'dearest Albert', Queen Victoria describes the wedding of her eldest son:

10 March 1863

All is over and this (to me) most trying day is past, as a dream, for all seems like a dream now and leaves hardly any impression upon my poor mind and broken heart! Here I sit lonely and desolate, who so need love and tenderness, while our two daughters have each their loving husbands, and Bertie has taken his lovely, pure, sweet Bride to Osborne, such a jewel whom he is indeed lucky to have obtained. How I pray God may ever bless them! Oh! what I suffered in the Chapel [St George's Chapel, Windsor], where all that was joy, pride, and happiness on January 25th, '58 [when the Princess Royal was married at St James's], was repeated without the principal figure of all, the guardian angel of the family, being there. It was indescribable. At one moment, when I first heard the flourish of trumpets, which brought back to my mind my whole life of twenty years at his dear side, safe, proud, secure, and happy, I felt as if I should faint. Only by a violent effort could I succeed in mastering my emotion!

When the celebrations were over:

We hastened to my room, where I saw them drive off, through the enthusiastic crowds. It was so like our driving away twenty-three years ago to Windsor, amidst the same crowds and shouts of joy! Aunt Cambridge and Mary came in to wish me good-bye, and then I drove with Lenchen down to the Mausoleum, and prayed by that beloved resting-place, feeling soothed and calmed.

QUEEN VICTORIA, from her Journal

The pain of loss and separation comes through Dorothy Wordsworth's restrained account of her brother's wedding:

On Monday 4th October 1802, my Brother William was married to Mary Hutchinson. I slept a good deal of the night and rose fresh and well in the morning. At a little after 8 o'clock I saw them go down the avenue towards the Church. William had parted from me upstairs. I gave him the wedding ring—with how deep a blessing! I took it from my forefinger where I had worn it the whole of the night before—he slipped it again onto my finger and blessed me fervently. When they were absent my dear little Sara prepared the breakfast. I kept myself as quiet as I could, but when I saw the two men running up the walk, coming to tell us it was over, I could stand it no longer and threw myself on the bed where I lay in stillness, neither hearing or seeing any thing, till Sara came upstairs to me and said 'They are coming'. This forced me from the bed where I lay and I moved I knew not how straight forward, faster than my strength could carry me till I met my beloved William and fell upon his bosom.

DOROTHY WORDSWORTH, from her Journal

Dorothy did, however, leave with William and Mary on their wedding journey, unlike Frankie in Carson McCullers's story:

The wedding was all wrong, although she could not point out single faults. . . . She wanted to speak to her brother and the bride, to talk to them and tell them of her plans, the three of them alone together. But they were never once alone; Jarvis was out checking the car someone was lending for the honeymoon, while Janice dressed in the front bedroom among a crowd of beautiful grown girls. She wandered from one to the other of them, unable to explain. And once Janice put her arms around her, and said she was so glad to have a little sister—and when Janice kissed her, F. Jasmine felt an aching in her throat and could not speak. Jarvis, when she went to find him in the yard, lifted her up in a roughhouse way and said: 'Frankie the lankie the alaga fankie, the tee-legged toe-legged bow-legged Frankie.' And he gave her a dollar.

She stood in the corner of the bride's room, wanting to say: I love the two of you so much and you are the we of me. Please take me with you from the wedding, for we belong to be together. Or even if she could have said: May I trouble you to step into the next room, as I have something to reveal to you and Jarvis? And get the three of them in a room alone together and somehow manage to explain. If only she had written it down on the typewriter in advance, so that she could hand it to them and they would read! But this she had not thought to do, and her tongue was heavy in her mouth and dumb.

*

The wedding was like a dream outside her power, or like a show unmanaged by her in which she was supposed to have no part. The living-room was crowded with Winter Hill company, and the bride and her brother stood before the mantelpiece at the end of the room. And seeing them again together was more like singing feeling than a picture that her dizzied eyes could truly see. She watched them with her heart, but all the time she was only thinking: I have not told them and they don't know. And knowing this was heavy as a swallowed stone. And afterwards, during the kissing of the bride, refreshments served in the dining-room, the stir and party bustle—she hovered close to the two of them, but words would not come. They are not going to take me, she was thinking, and this was the one thought she could not bear.

When Mr Williams brought their bags, she hastened after with her own suitcase. The rest was like some nightmare show in which a wild girl in the audience breaks on to the stage to take upon herself an unplanned part that was never written or meant to be. You are the we of me, her heart was saying, but she could only say aloud: 'Take me!' And they pleaded and begged with her, but she was already in the car. At the last she clung to the steering wheel until her father and somebody else had hauled and dragged her from the car, and even then she could only cry in the dust of the empty road: 'Take me! Take me!' But there was only the wedding company to hear, for the bride and her brother had driven away.

CARSON MCCULLERS, *The Member of the Wedding*, 1946

Sophie came at seven to dress me; she was very long indeed in accomplishing her task; so long, that Mr Rochester, grown, I suppose, impatient of my delay, sent up to ask why I did not come. She was just fastening my veil (the plain square of blond after all) to my hair with a brooch; I hurried from under her hands as soon as I could.

'Stop!' she cried in French. 'Look at yourself in the mirror: you have not taken one peep.'

So I turned at the door: I saw a robed and veiled figure, so unlike my usual self that it seemed almost the image of a stranger. 'Jane!' called a voice, and I hastened down. I was received at the foot of the stairs by Mr Rochester.

'Lingerer,' he said, 'my brain is on fire with impatience; and you tarry so long!'

*

Our place was taken at the communion rails. Hearing a cautious step behind me, I glanced over my shoulder: one of the strangers—a gentleman, evidently—was advancing up the chancel. The service began. The explanation of the intent of matrimony was gone through; and then the clergyman came a step further forward, and, bending slightly towards Mr Rochester, went on:—

'I require and charge you both (as ye will answer at the dreadful day of Judgment, when the secrets of all hearts shall be disclosed), that if either of you know any impediment why ye may not lawfully be joined together in matrimony ye do now confess it; for be ye well assured that so many as are coupled together otherwise than God's Word doth allow, are not joined together by God, neither is their matrimony lawful.'

He paused, as the custom is. When is the pause after that sentence ever broken by reply? Not, perhaps, once in a hundred years. And the clergyman, who had not lifted his eyes from his book, and had held his breath but for a moment, was proceeding: his hand was already stretched towards Mr Rochester, as his lips unclosed to ask, 'Wilt thou have this woman for thy wedded wife?'—when a distinct and near voice said:—

'The marriage cannot go on; I declare the existence of an impediment.'

The clergyman looked up at the speaker, and stood mute; the clerk did the same; Mr Rochester moved slightly, as if an earthquake had rolled under his feet: taking a firmer footing, and not turning his head or eyes, he said, 'Proceed.'

Profound silence fell when he had uttered that word, with deep but low intonation. Presently Mr Wood said:—

'I cannot proceed without some investigation into what has been asserted, and evidence of its truth or falsehood.'

'The ceremony is quite broken off,' subjoined the voice behind us. 'I am in a condition to prove my allegation: an insuperable impediment to this marriage exists.'

Mr Rochester heard, but heeded not; he stood stubborn and rigid: making no movement but to possess himself of my hand. What a hot and strong grasp he had! and how like quarried marble was his pale, firm, massive front at this moment! How his eye shone, still, watchful, and yet wild beneath!

Mr Wood seemed at a loss. 'What is the nature of the impediment?' he asked. 'Perhaps it may be got over—explained away?'

'Hardly,' was the answer; 'I have called it insuperable, and I speak advisedly.'

The speaker came forwards, and leaned on the rails. He continued, uttering each word distinctly, calmly, steadily, but not loudly—

'It simply consists in the existence of a previous marriage. Mr Rochester has a wife now living.'

My nerves vibrated to those low-spoken words as they had never vibrated to thunder—my blood felt their subtle violence as it had never felt frost or fire: but I was collected, and in no danger of swooning. I looked at Mr Rochester: I made him look at me. His whole face was colourless rock: his eye was both spark and flint. He disavowed nothing: he seemed as if he would defy all things. Without speaking; without smiling; without seeming to recognize in me a human being, he only twined my waist with his arm, and riveted me to his side.

'Who are you?' he asked of the intruder.

'My name is Briggs, a solicitor, of —— Street, London.'

'And you would thrust on me a wife?'

'I would remind you of your lady's existence, sir; which the law recognizes, if you do not.'

'Favour me with an account of her—with her name, her parentage, her place of abode.'

'Certainly.' Mr Briggs calmly took a paper from his pocket, and read out in a sort of official nasal voice,—

' "I affirm and can prove that on the 20th of October, AD—, (a date of fifteen years back), Edward Fairfax Rochester, of Thornfield Hall, in the county of ——, and of Ferndean Manor, in ——shire, England, was married to my sister, Bertha Antoinetta Mason, daughter of Jonas Mason, merchant, and of Antoinetta his wife, a Creole, at —— church, Spanish Town, Jamaica. The record of the marriage will be found in the register of that church—a copy of it is now in my possession. Signed, Richard Mason." '

'That—if a genuine document—may prove I have been married, but it does not prove that the woman mentioned therein as my wife is still living.'

'She was living three months ago,' returned the lawyer.

'How do you know?'

'I have a witness to the fact, whose testimony even you, sir, will scarcely controvert.'

'Produce him—or go to hell.'

'I will produce him first—he is on the spot. Mr Mason, have the goodness to step forward.'

Mr Rochester, on hearing the name, set his teeth; he experienced, too, a sort of strong convulsive quiver; near to him as I was, I felt the spasmodic movement of fury or despair run through his frame. The second stranger, who had hitherto lingered in the background, now drew near; a pale face looked over the solicitor's shoulder—yes, it was Mason himself. Mr Rochester turned and glared at him. His eye, as I have often said, was a black eye: it had now a tawny, nay, a bloody light in its gloom; and his face flushed—olive cheeks and hueless forehead received a glow as from spreading, ascending heart-fire; and he stirred, lifted his strong arm—he could have struck Mason, dashed him on the church floor, shocked by ruthless blow the breath from his body—but Mason shrank away, and cried faintly, 'Good God!' Contempt fell cool on Mr Rochester—his passion died as if a blight had shrivelled it up: he only asked, 'What have *you* to say?'

An inaudible reply escaped Mason's white lips.

'The devil is in it if you cannot answer distinctly. I again demand what have *you* to say?'

'Sir—sir,' interrupted the clergyman, 'do not forget you are in a sacred place.' Then addressing Mason, he inquired gently. 'Are you aware, sir, whether or not this gentleman's wife is still living?'

'Courage,' urged the lawyer; 'speak out.'

'She is now at Thornfield Hall,' said Mason, in more articulate tones: 'I saw her there last April. I am her brother.'

'At Thornfield Hall!' ejaculated the clergyman. 'Impossible! I am an old resident in this neighbourhood, sir, and I never heard of a Mrs Rochester at Thornfield Hall.'

I saw a grim smile contort Mr Rochester's lips, and he muttered—

'No, by God! I took care that none should hear of it—or of her under that name.' He mused—for ten minutes he held counsel with himself: he formed his resolve and announced it.

'Enough! all shall bolt out at once, like the bullet from the barrel.

Wood, close your book, and take off your surplice; John Green (to the clerk), leave the church: there will be no wedding to-day.'

CHARLOTTE BRONTË, *Jane Eyre*, 1847

John Osborne's anti-hero, prototype of the 'angry young man', taunts his wife with a description of their wedding:

JIMMY: The last time she was in a church was when she was married to me. I expect that surprises you, doesn't it? It was expediency, pure and simple. We were in a hurry, you see. (*The comedy of this strikes him at once, and he laughs.*) Yes, we were actually in a hurry! Lusting for the slaughter! Well, the local registrar was a particular pal of Daddy's, and we knew he'd spill the beans to the Colonel like a shot. So we had to seek out some local vicar who didn't know him quite so well. But it was no use. When my best man—a chap I'd met in the pub that morning—and I turned up, Mummy and Daddy were in the church already. They'd found out at the last moment, and had come to watch the execution carried out. How I remember looking down at them, full of beer for breakfast, and feeling a bit buzzed. Mummy was slumped over her pew in a heap—the noble, female rhino, pole-axed at last! And Daddy sat beside her, upright and unafraid, dreaming of his days among the Indian Princes, and unable to believe he'd left his horsewhip at home. Just the two of them in that empty church—them and me. (*Coming out of his remembrance suddenly.*) I'm not sure what happened after that. We must have been married, I suppose. I think I remember being sick in the vestry. (*To Alison.*) Was I?

JOHN OSBORNE, *Look Back in Anger*, 1956

It was June 1933, one week after Commencement, when Kay Leiland Strong, Vassar '33, the first of her class to run around the table at the Class Day dinner, was married to Harald Petersen, Reed '27, in the chapel of St George's Church, P.E., Karl F. Reiland, Rector. Outside, on Stuyvesant Square, the trees were in full leaf, and the wedding guests arriving by twos and threes in taxis heard the voices of children playing round the statue of Peter Stuyvesant in the park. Paying the driver, smoothing out their gloves, the pairs and trios of young women, Kay's classmates, stared about them curiously, as though they were in a foreign city.

*

The sense of an adventure was strong on them this morning, as they seated themselves softly in the still, near-empty chapel; they had never been to a wedding quite like this one before, to which invitations had been issued orally by the bride herself, without the intervention of a relation or any older person, friend of the family. There was to be no honeymoon, they had heard, because Harald (that was the way he spelled it—the old Scandinavian way) was working as an assistant stage manager for a theatrical production and had to be at the theatre as usual this evening to call 'half hour' for the actors. This seemed to them very exciting and of course it justified the oddities of the wedding: Kay and Harald were too busy and dynamic to let convention cramp their style.

*

They had pondered about Harald; Kay had met him last summer when she was working as an apprentice at a summer theatre in Stamford and both sexes had lived in a dormitory together. She said he wanted to marry her, but that was not the way his letters sounded to the group. They were not love letters at all, so far as the group could see, but accounts of personal successes among theatrical celebrities, what Edna Ferber had said to George Kaufman in his hearing, how Gilbert Miller had sent for him and a woman star had begged him to read his play to her in bed. 'Consider yourself kissed,' they ended, curtly, or just 'CYK'—not another word. In a young man of their own background, as the girls vaguely phrased it, such letters would have been offensive, but their education had impressed on them the unwisdom of making large judgments from one's own narrow little segment of experience. Still, they could tell that Kay was not as sure of him as she pretended she was; sometimes he did not write for weeks, while poor Kay went on whistling in the dark. . . . Then, that night, when Kay had run around the long table, which meant you were announcing your engagement to the whole class, and produced from her winded bosom a funny Mexican silver ring to prove it, their alarm had dissolved into a docile amusement; they clapped, dimpling and twinkling, with an air of prior knowledge. More gravely, in low posh tones, they assured their parents, up for the Commencement ceremonies, that the engagement was of long standing, that Harald was 'terribly nice' and 'terribly in love' with Kay. Now, in the chapel, they rearranged their fur pieces and smiled at each other, noddingly, like mature little martens and sables: they had been right, the hardness was only a phase; it was certainly a point for *their* side that the iconoclast and scoffer was the first of the little band to get married.

'Who would have thunk it?' irrepressibly remarked 'Pokey' (Mary) Prothero, a fat cheerful New York society girl with big red cheeks and yellow hair, who talked like a jolly beau of the McKinley period, in imitation of her yachtsman father.

*

'What perfect pets they look!' murmured Dottie Renfrew, of Boston, . . . as Harald and Kay came in from the vestry and took their places before the surpliced curate, accompanied by little Helena Davison, Kay's ex-room-mate from Cleveland, and by a sallow blond young man with a moustache. Pokey made use of her *lorgnon*, squinting up her pale-lashed eyes like an old woman; this was her first appraisal of Harald, for she had been away hunting for the week-end the one time he had come to college. 'Not too bad,' she pronounced. 'Except for the shoes.' . . . '*Shut up*,' came a furious growl from her other side. Pokey, hurt, peered around, to find Elinor Eastlake, of Lake Forest, the taciturn brunette beauty of the group, staring at her with murder in her long, green eyes.

*

To Elinor, this wedding was torture. Everything was so jaggedly ill-at-ease: Kay's costume, Harald's shoes and necktie, the bare altar, the sparsity of guests on the groom's side (a couple and a solitary man), the absence of any family connection. Intelligent and morbidly sensitive, she was inwardly screaming with pity for the principals and vicarious mortification. Hypocrisy was the sole explanation she could find for the antiphonal bird twitter of 'Terribly nice', and 'Isn't this exciting?' that had risen to greet the couple in lieu of a wedding march.

*

In a private dining-room of the hotel, Kay and Harald stood on a faded flowered carpet, receiving their friends' congratulations. A punch was being served, over which the guests were exclaiming: 'What *is* it?', 'Perfectly *delicious*', 'How did you ever think of it?' and so on. To each one, Kay gave the recipe. . . . The recipe was an ice-breaker—just as Kay had hoped, she explained aside to Helena Davison: everyone tasted it and agreed that it was the maple syrup that made all the difference.

*

'Isn't this fun?' cried Libby MacAusland, arching her long neck and weaving her head about and laughing in her exhausted style. '*So* much nicer,' purred the voices. 'No receiving line, no formality, no older people.' 'It's just what I want for myself,' announced Libby. 'A young people's wedding!' She uttered a blissful scream as a Baked Alaska came in, the meringue faintly smoking. 'Baked Alaska!' she cried and fell back, as if in a heap, on her chair. 'Girls!' she said solemnly, pointing to the big ice-cream cake with slightly scorched peaks of meringue that was being lowered into position before Kay. 'Look at it. Childhood dreams come true! It's every children's party in the whole blessed United States. It's patent-leather slippers and organdie and a shy little boy in an Eton collar asking you to dance. I don't know when I've been so excited. I haven't seen one since I was twelve years old. It's Mount Whitney; it's Fujiyama.'

*

The gabble of voices slowly died down. The girls, confused by alcohol, cast inquiring looks at each other. What was to happen now? At an ordinary wedding, Kay and Harald would slip off to change to travelling costume, and Kay would throw her bouquet. But there was to be no honeymoon, they recalled. Kay and Harald, evidently, had nowhere to go but back to the sublet apartment they had just left this morning. Probably, if the group knew Kay, the bed was not even made. The funny, uneasy feeling that had come over them all in the chapel affected them again. They looked at their watches; it was only one-fifteen. How many hours till it was time for Harald to go to work? Doubtless, lots of couples got married and just went home again, but somehow it did not seem right to let that happen.

*

Pokey Prothero's voice, like a querulous grackle, intervened. 'You two are supposed to go away,' she suddenly complained, crushing out her cigarette and looking through her *lorgnon* with an air of surprised injury from the bride to the groom. Trust Pokey, thought the girls, with a joint sigh. 'Where should we go, Pokey?' answered Kay, smiling. 'Yes, Pokey, where should we go?' agreed the bridegroom. Pokey considered. 'Go to Coney Island,' she said. Her tone of irrefutable, self-evident logic, like that of an old man or a child, took everyone aback for a second. 'What a splendid idea!' cried Kay. 'On the subway?' 'Brighton Express, via Flatbush Avenue,' intoned Harald. 'Change at Fulton Street.' 'Pokey, you're a genius,' said everyone, in voices of immense relief.

*

'Kay must throw her bouquet!' shrieked Libby MacAusland, stretching on her long legs, like a basketball centre, as a crowd of people massed to watch them. 'My girl's from Vassar; none can surpass 'er,' the radio man struck up. Harald produced two nickels and the newlyweds passed through the turnstile; Kay, who, all agreed, had never looked prettier, turned and threw her bouquet, high in the air, back over the turnstiles to the waiting girls. Libby jumped and caught it, though it had really been aimed at Priss just behind her. And at that moment Lakey gave them all a surprise; the brown-paper parcels she had checked in the hotel proved to contain rice. '*That* was what you stopped for!' exclaimed Dottie, full of wonder, as the wedding party seized handfuls and pelted them after the bride and the groom; the platform was showered with white grains when the local train finally came in. 'That's banal! That's not like you, Eastlake!' Kay turned and shouted as the train doors were closing, and everyone, dispersing, agreed that it was not like Lakey at all, but that, banal or not, it was just the little touch that had been needed to round off an unforgettable occasion.

MARY McCARTHY, *The Group*, 1954

He knew well, too well, in the secret centre of his brain, that Arabella was not worth a great deal as a specimen of womankind. Yet, such being the custom of the rural districts among honourable young men who had drifted so far into intimacy with a woman as he unfortunately had done, he was ready to abide by what he had said, and take the consequences. For his own soothing he kept up a factitious belief in her. His idea of her was the thing of most consequence, not Arabella herself, he sometimes said laconically.

The banns were put in and published the very next Sunday. The people of the parish all said what a simple fool young Fawley was. All his reading had only come to this, that he would have to sell his books to buy saucepans. Those who guessed the probable state of affairs, Arabella's parents being among them, declared that it was the sort of conduct they would have expected of such an honest young man as Jude in reparation of the wrong he had done his innocent sweetheart. The parson who married them seemed to think it satisfactory too.

And so, standing before the aforesaid officiator, the two swore that at every other time of their lives till death took them, they would assuredly believe, feel, and desire precisely as they had believed, felt, and desired during the few preceding weeks. What was as remarkable as the undertaking itself was the fact that nobody seemed at all surprised at what they swore.

THOMAS HARDY, *Jude the Obscure*, 1895

MARRYING ABSURD

Driving in across the Mojave from Los Angeles, one sees the signs way out on the desert, looming up from that moonscape of rattlesnakes and mesquite, even before the Las Vegas lights appear like a mirage on the horizon: 'GETTING MARRIED? Free License Information First Strip Exit.' Perhaps the Las Vegas wedding industry achieved its peak operational efficiency between 9:00 p.m. and midnight of August 26, 1965, an otherwise unremarkable Thursday which happened to be, by Presidential order, the last day on which anyone could improve his draft status merely by getting married. One hundred and seventy-one couples were pronounced man and wife in the name of Clark County and the State of Nevada that night, sixty-seven of them by a single justice of the peace, Mr James A. Brennan. Mr Brennan did one wedding at the Dunes and the other sixty-six in his office, and charged each couple eight dollars. One bride lent her veil to six others. 'I got it down from five to three minutes,' Mr Brennan said later of his feat. 'I could've married them *en masse*, but they're people, not cattle. People expect more when they get married.'

What people who get married in Las Vegas actually do expect—what, in the largest sense, their 'expectations' are—strikes one as a curious and self-contradictory business. Las Vegas is the most extreme and allegorical of American settlements, bizarre and beautiful in its venality and in its devotion to immediate gratification, a place the tone of which is set by mobsters and call girls and ladies' room attendants with amyl nitrite poppers in their uniform pockets. Almost everyone notes that there is no 'time' in Las Vegas, no night and no day and no past and no future (no Las Vegas casino, however, has taken the obliteration of the ordinary true sense quite so far as Harold's Club in Reno, which for a while issued, at odd intervals in the day and night, mimeographed 'bulletins' carrying news from the world outside); neither is there any logical sense of where one is. One is standing on a highway in the middle of a vast hostile desert looking at an eighty-foot sign which blinks 'STARDUST' or 'CAESAR'S PALACE.' Yes, but what does that explain? This geographical implausibility reinforces the sense that what happens there has no connection with 'real' life; Nevada cities like Reno and Carson are ranch towns, Western towns, places behind which there is some historical imperative. But Las Vegas seems to exist only in the eye of the beholder. All of which makes it an extraordinary stimulating and interesting place, but an odd one in which to want to wear a candlelight satin Priscilla of Boston wedding dress with Chantilly lace insets, tapered sleeves and a detachable modified train.

And yet the Las Vegas wedding business seems to appeal to precisely that impulse. 'Sincere and Dignified Since 1954,' one wedding chapel advertises. There are nineteen such wedding chapels in Las Vegas, intensely competitive, each offering better, faster, and, by implication, more sincere services than the next: Our Photos Best Anywhere, Your Wedding on A Phonograph Record, Candlelight with Your Ceremony, Honeymoon Accommodations, Free Transportation from Your Motel to Courthouse to Chapel and Return to Motel, Religious or Civil Ceremonies, Dressing Rooms, Flowers, Rings, Announcements, Witnesses Available, and Ample Parking. All of these services, like most others in Las Vegas (sauna baths, payroll-check cashing, chinchilla coats for sale or rent) are offered twenty-four hours a day, seven days a week, presumably on the premise that marriage, like craps, is a game to be played when the table seems hot.

But what strikes one most about the Strip chapels, with their wishing wells and stained-glass paper windows and their artificial bouvardia, is that so much of their business is by no means a matter of simple convenience, of late-night liaisons between show girls and baby Crosbys. Of course there is some of that. (One night about eleven o'clock in Las Vegas I watched a bride in an orange minidress and masses of flame-colored hair stumble from a Strip chapel on the arm of her bridegroom, who looked the part of the expendable nephew in movies like *Miami Syndicate*. 'I gotta get the kids,' the bride whimpered. 'I gotta pick up the sitter, I gotta get to the midnight show.' 'What you gotta get,' the bridegroom said, opening the door of a Cadillac Coupe de Ville and watching her crumple on the seat, 'is sober.') But Las Vegas seems to offer something other than 'convenience'; it is merchandising 'niceness,' the facsimile of proper ritual, to children who do not know how else to find it, how to make the arrangements, how to do it 'right.' All day and evening long on the Strip, one sees actual wedding parties, waiting under the harsh lights at a crosswalk, standing uneasily in the parking lot of the Frontier while the photographer hired by The Little Church of the West ('Wedding Place of the Stars') certifies the occasion, takes the picture: the bride in a veil and white satin pumps, the bridegroom usually in a white dinner jacket, and even an attendant or two, a sister or a best friend in hot-pink *peau de soie*, a flirtation veil, a carnation nosegay. 'When I Fall in Love It Will Be Forever,' the organist plays, and then a few bars of Lohengrin. The mother cries; the stepfather, awkward in his role, invites the chapel hostess to join them for a drink at the Sands. The hostess declines with a professional smile; she has already transferred her interest to the group waiting outside. One bride out, another in, and again the sign goes up on the chapel door: 'One moment please—Wedding.'

I sat next to one such wedding party in a Strip restaurant the last time I was in Las Vegas. The marriage had just taken place; the bride still wore her dress, the mother her corsage. A bored waiter poured out a few swallows of pink champagne ('on the house') for everyone but the bride, who was too young to be served. 'You'll need something with more kick than that,' the bride's father said with heavy jocularity to his new son-in-law; the ritual jokes about the wedding night had a certain Panglossian character, since the bride was clearly several months pregnant. Another round of pink champagne, this time not on the house, and the bride began to cry. 'It was just as nice,' she sobbed, 'as I hoped and dreamed it would be.'

JOAN DIDION, *Slouching towards Bethlehem*, 1961

THE WHITSUN WEDDINGS

That Whitsun, I was late getting away:
  Not till about
One-twenty on the sunlit Saturday
Did my three-quarters-empty train pull out,
All windows down, all cushions hot, all sense
Of being in a hurry gone. We ran
Behind the backs of houses, crossed a street
Of blinding windscreens, smelt the fish-dock; thence
The river's level drifting breadth began,
Where sky and Lincolnshire and water meet.

All afternoon, through the tall heat that slept
  For miles inland,
A slow and stopping curve southwards we kept.
Wide farms went by, short-shadowed cattle, and
Canals with floatings of industrial froth;
A hothouse flashed uniquely: hedges dipped
And rose: and now and then a smell of grass
Displaced the reek of buttoned carriage-cloth
Until the next town, new and nondescript,
Approached with acres of dismantled cars.

At first, I didn't notice what a noise
  The weddings made
Each station that we stopped at: sun destroys
The interest of what's happening in the shade,
And down the long cool platforms whoops and skirls
I took for porters larking with the mails,
And went on reading. Once we started, though,
We passed them, grinning and pomaded, girls
In parodies of fashion, heels and veils,
All posed irresolutely, watching us go,

As if out on the end of an event
  Waving goodbye
To something that survived it. Struck, I leant
More promptly out next time, more curiously,
And saw it all again in different terms:

The fathers with broad belts under their suits
And seamy foreheads; mothers loud and fat;
An uncle shouting smut; and then the perms,
The nylon gloves and jewellery-substitutes,
The lemons, mauves, and olive-ochres that

Marked off the girls unreally from the rest.
  Yes, from cafés
And banquet-halls up yards, and bunting-dressed
Coach-party annexes, the wedding-days
Were coming to an end. All down the line
Fresh couples climbed aboard: the rest stood round;
The last confetti and advice were thrown,
And, as we moved, each face seemed to define
Just what it saw departing: children frowned
At something dull; fathers had never known

Success so huge and wholly farcical;
  The women shared
The secret like a happy funeral;
While girls, gripping their handbags tighter, stared
At a religious wounding. Free at last,
And loaded with the sum of all they saw,
We hurried towards London, shuffling gouts of steam.
Now fields were building-plots, and poplars cast
Long shadows over major roads, and for
Some fifty minutes, that in time would seem

Just long enough to settle hats and say
  I nearly died,
A dozen marriages got under way.
They watched the landscape, sitting side by side
—An Odeon went past, a cooling tower,
And someone running up to bowl—and none
Thought of the others they would never meet
Or how their lives would all contain this hour.
I thought of London spread out in the sun,
Its postal districts packed like squares of wheat:

There we were aimed. And as we raced across
  Bright knots of rail
Past standing Pullmans, walls of blackened moss
Came close, and it was nearly done, this frail
Travelling coincidence; and what it held
Stood ready to be loosed with all the power
That being changed can give. We slowed again,
And as the tightened brakes took hold, there swelled
A sense of falling, like an arrow-shower
Sent out of sight, somewhere becoming rain.

PHILIP LARKIN, 1958

4
From this Day Forward

The Owl and the Pussy-Cat went to sea
 In a beautiful pea-green boat,
They took some honey, and plenty of money,
 Wrapped up in a five-pound note.
The Owl looked up to the stars above,
 And sang to a small guitar,
'O lovely Pussy! O Pussy, my love,
 'What a beautiful Pussy you are,
  'You are,
  'You are!
'What a beautiful Pussy you are!'

EDWARD Lear evokes the insouciant way in which many young couples set out on their voyage into life together, with scant provision on board apart from the sweet honey of love, full of romantic idealization of their partner and little practical thought for what lies ahead.

To fall in love is to invest the loved one with all the characteristics and qualities that consciously and unconsciously we seek in a partner. The process and the task of marriage is to come to terms with the reality.

If the marriage has been based only on romantic fantasy, then the fantasist, like Flaubert's Emma Bovary, will soon be disappointed. Or if, like George Eliot's Dorothea in *Middlemarch*, one partner has projected on to the other 'every quality that she herself brought', disillusion will be fought against and be all the more painful. Dorothea's struggle to make her marriage into her image of what such a union should and could be, is central to this chapter.

'Seldom or never does a marriage develop into an individual relationship smoothly and without crises. There is no birth of consciousness without pain' (C. G. Jung). Even in such idyllic partnerships as Levin and Kitty's in Tolstoy's *Anna Karenina*, adjustments have to be made, and Levin has to accept that his 'poetic, exquisite' Kitty has a mundane and practical side too, and 'can fuss about table-cloths, furniture, spare-room mattresses, a tray, the cook, the dinner and so forth'. In their relationship, as in Natasha and Pierre's in *War and Peace*, the reality is not too far removed from the idealized fantasy, and both these couples are able to continue to meet each other's changing needs. Ironically, Tolstoy, whose own marriage became so devastatingly unhappy, has given us a picture of two marriages of profound happiness, where each partner glows with the heightened self-esteem that such a mutually affirming relationship can give.

> After seven years of married life Pierre was able to feel a comforting and assured conviction that he was not a bad fellow after all. This he could do because he saw himself mirrored in his wife. In himself he felt the good and bad inextricably mixed and overlapping. But in his wife he saw reflected only what was really good in him, since everything else she rejected. And this reflection was not the result of a logical process of thought but came from some other mysterious, direct source.

Though most of the authors quoted in this chapter are men, women's dissatisfaction with the institution of marriage forms a keynote. 'Is this all?'—the clarion call with which Betty Friedan brought the mid-twentieth-century feminist movement to life, echoes the cry of women through the ages, from Euripides' Medea, through Ibsen's Nora, whose 'slammed door', according to a contemporary critic, 'reverberated across the roof of the world', to the present day. Do women expect more from their emotional life than men do, or are they merely seen to be more self-analytical and more articulate about their feelings?

This chapter contains little poetry, for it concerns itself chiefly with the prose of married life.

A young husband and wife came to stay with us in all the first flush of married happiness. One realised all day long that other people merely made a pleasant background for their love, and that for each there was but one real figure on the scene. This was borne witness to by a whole armoury of gentle looks, swift glances, silent gestures. They were both full to the brim of a delicate laughter, of over-brimming wonder, of tranquil desire. And we all took a part in their gracious happiness. In the evening they sang and played to us, the wife being an accomplished pianist, the husband a fine singer. But though the glory of their art fell in rainbow showers on the audience, it was for each other that they sang and played.

These two spirits seemed, with hands intertwined, to have ascended gladly into the mountain, and to have seen a transfiguration of life: which left them not in a blissful eminence of isolation, but rather, as it were, beckoning others upwards, and saying that the road was indeed easy and plain. And so the sweet hour passed and left a fragrance behind it; whatever might befall, they had tasted of the holy wine of joy, they had blessed the cup, and bidden us, too, to set our lips to it.

A. C. BENSON (1862–1925)

Unconsciousness results in non-differentiation, or unconscious identity. The practical consequence of this is that one person presupposes in the other a psychological structure similar to his own. Normal sex life, as a shared experience with apparently similar aims, further strengthens the feeling of unity and identity. This state is described as one of complete harmony, and is extolled as a great happiness ('one heart and one soul')—not without good reason, since the return to that original condition of unconscious oneness is like a return to childhood. Hence the childish gestures of all lovers. Even more is it a return to the mother's womb, into the teeming depths of an as yet unconscious creativity. It is, in truth, a genuine and incontestable experience of the Divine, whose transcendent force obliterates and consumes everything individual; a real communion with life and the impersonal power of fate. The individual will for self-possession is broken: the woman becomes the mother, the man the father, and thus both are robbed of their freedom and made instruments of the life urge.

C. G. JUNG, 'Marriage as a Psychological Relationship', 1925

'I don't intend him, or any man or any woman to be all my life—good heavens, no! There are heaps of things in me that he doesn't, and never shall, understand.'

Thus she spoke before the wedding ceremony and the physical union, before the astonishing glass shade had fallen that interposes between married couples and the world. She was to keep her independence more than do most women as yet. Marriage was to alter her fortunes rather than her character, and she was not far wrong in boasting that she understood her future husband. Yet he did alter her character—a little. There was an unforeseen surprise, a cessation of the winds and odours of life, a social pressure that would have her think conjugally.

E. M. FORSTER, *Howards End*, 1910

A BACHELOR'S COMPLAINT OF THE BEHAVIOUR OF MARRIED PEOPLE

As a single man, I have spent a good deal of my time in noting down the infirmities of Married People, to console myself for those superior pleasures, which they tell me I have lost by remaining as I am.

I cannot say that the quarrels of men and their wives ever made any great impression upon me, or had much tendency to strengthen me in those anti-social resolutions, which I took up long ago upon more substantial considerations. What oftenest offends me at the houses of married persons where I visit, is an error of quite a different description;—it is that they are too loving.

Not too loving neither: that does not explain my meaning. Besides, why should that offend me? The very act of separating themselves from the rest of the world, to have the fuller enjoyment of each other's society, implies that they prefer one another to all the world.

But what I complain of is, that they carry this preference so undisguisedly, they perk it up in the faces of us single people so shamelessly, you cannot be in their company a moment without being made to feel, by some indirect hint or open avowal, that *you* are not the object of this preference.

CHARLES LAMB, *Essays of Elia*, 1823

After all, the rosy love-making and marrying and Epithalamy are no more than the dawn of things, and to follow comes all the spacious interval of white laborious light. Try as we may to stay those delightful moments they fade and pass remorselessly; there is no returning, no recovering, only—for the foolish—the vilest peep-shows and imitations in dens and darkened rooms. We go on—we grow. At least we age. Our young couple, emerging presently from an atmosphere of dusk and morning stars, found the sky gathering grayly overhead and saw one another for the first time clearly in the light of every day.

H. G. WELLS, *Love and Mr Lewisham*, 1900

This is just about when she discovers that love and marriage mean a different thing for a male than they do for her: though men in general believe women in general to be inferior, every man has reserved a special place in his mind for the one woman he will elevate above the rest by virtue of association with himself. Until now the woman, out in the cold, begged for his approval, dying to clamber onto this clean well-lighted place. But once there, she realizes that she was elevated above other women not in recognition of her real value, but only because she matched nicely his store-bought pedestal. Probably he doesn't even know who she is (if indeed by this time she herself knows). He has let her in not because he genuinely loved her, but only because she played so well into his preconceived fantasies. Though she knew his love to be false, since she herself engineered it, she can't help feeling contempt for him. But she is afraid, at first, to reveal her true self, for then perhaps even that false love would go. And finally she understands that for him, too, marriage had all kinds of motivations that had nothing to do with love. She was merely the one closest to his fantasy image: she has been named Most Versatile Actress for the multi-role of Alter Ego, Mother of My Children, Housekeeper, Cook, Companion, in *his* play. She has been bought to fill an empty space in his life; but her life is nothing.

SHULAMITH FIRESTONE, *The Dialectics of Sex*, 1979

HARRIET AND LOVAT AT SEA IN MARRIAGE

When a sincere man marries a wife, he has one or two courses open to him, which he can pursue with that wife. He can propose to himself to be (*a*) the lord and master who is honoured and obeyed, (*b*) the perfect lover, (*c*) the true friend and companion. Of these (*a*) is now rather out of date. The lord and master has been proved, by most women quite satisfactorily, to be no more than a grown-up child, and his arrogance is to be tolerated just as a little boy's arrogance is tolerated, because it is rather amusing, and up to a certain point becoming. The case of (*b*), the perfect lover, is the crux of all ideal marriage today. But alas, not even the lord and master turns out such a fiasco as does the perfect lover, ninety-nine times out of a hundred. The perfect lover marriage ends usually in a quite ghastly anti-climax, divorce and horrors and the basest vituperation. Alas for the fact, as compared with the ideal. A marriage of the perfect-lover type is bound either to end in catastrophe, or to slide away towards (*a*) or (*c*). It must either revert to a mild form of the lord-and-master marriage, and a wise woman, who knows the sickeningness of catastrophes and the ridiculous futility of second shots at the perfect-love paradise, often wisely pushes the marriage back gradually into one of the little bays or creeks of this Pacific ocean of marriage, lord-and-masterdom. Not that either party really believes in the lordship of man. But you've got to get into still water some time or other. The perfect-love business inevitably turns out to be a wildly stormy strait, like the Straits of Magellan, where two fierce and opposing currents meet and there is the devil of a business trying to keep the bark of marriage, with the flag of perfect love at the mast, from dashing on a rock or foundering in the heavy seas. Two fierce and opposing currents meet in the narrows of perfect love. They may meet in blue and perfect weather, when the albatross hovers in the great sky like a permanent benediction, and the sea shimmers a second heaven. But you needn't wait long. The seas will soon begin to rise, the ship to roll. And the waters of perfect love—when once this love is consummated in marriage—become inevitably a perfect hell of storms and furies.

Then, as I say, the hymeneal bark either founders, or dashes on a rock, or more wisely gets out of the clash of meeting oceans and takes one tide or the other, where the flood has things all its own way. The woman, being today the captain of the marriage bark, either steers into the vast Pacific waters of lord-and-masterdom, though never, of course, hauling down the flag of perfect love; or else, much more frequently these latter days, she steers into the rather grey Atlantic of true friendship and companionship, still keeping the flag of perfect love bravely afloat.

And now the bark is fairly safe. In the great Pacific, the woman can take the ease and warm repose of her new dependence, but she is usually laughing up her sleeve. She lets the lord and master manage the ship, but woe betide him if he seeks to haul down the flag of perfect love. There is mutiny in a moment. And his chief officers and his crew, namely his children and his household servants, are up and ready to put him in irons at once, at a word from that wondrous goddess of the bark, the wife of his bosom. It is Aphrodite, mistress of the seas, in her grand capacity of motherhood and attendant wifehood. None the less, with a bit of managing the hymeneal bark sails on across the great waters into port. A lord and master is not much more than an upper servant while the flag of perfect love is flying and the sea-mother is on board. But a servant with the name of captain, and the pleasant job of sailing the ship and giving the necessary orders. He feels it is quite all right. He is supreme servant-in-command, while the mistress of mistresses smiles as she suckles his children. She is suckling him too.

D. H. LAWRENCE, *Kangaroo*, 1923

And yet sometimes it occurred to her that this was the finest time of her life, the so-called honeymoon. To savour all its sweetness, it would doubtless have been necessary to sail away to lands with musical names where wedding nights leave behind them a more delicious indolence. In a post-chaise, behind blue silk blinds, you climb at a foot-pace up precipitous roads, listening to the postillion's song echoing across the mountain, amid the tinkling of goat-bells and the muffled noise of waterfalls. At sunset you breathe the scent of lemon trees on the shore of a bay. At night, together on the terrace of your villa, with fingers intertwined, you gaze at the stars and make plans for the future. It seemed to her that certain parts of the world must produce happiness, as they produce peculiar plants which will flourish nowhere else. Why could she not now be leaning on the balcony of a Swiss chalet, or immuring her sadness in a Scotch cottage, with a husband in a black velvet coat with long flaps, and soft boots, and peaked hat, and ruffles!

She would have been glad of someone in whom to confide all this; but how describe an intangible unease, that shifts like the clouds and eddies like the wind? Lacking the words, she had neither the opportunity nor the courage.

Nevertheless, had Charles so wished, had he guessed, had his eyes once read her thoughts, it would instantly have delivered her heart of a rich load, as a single touch will bring the ripe fruit falling from the tree. But as their outward familiarity grew, she began to be inwardly detached, to hold herself more aloof from him.

Charles' conversation was as flat as a street pavement, on which everybody's ideas trudged past, in their workaday dress, provoking no emotion, no laughter, no dreams. At Rouen, he said, he had never had any desire to go and see a Paris company at the theatre. He couldn't swim, or fence, or fire a pistol, and was unable to explain a riding term she came across in a novel one day.

Whereas a man, surely, should know about everything; excel in a multitude of activities, introduce you to passion in all its force, to life in all its grace, initiate you into all mysteries! But this one had nothing to teach; knew nothing, wanted nothing. He thought she was happy; and she hated him for that placid immobility, that stolid serenity of his, for that very happiness which she herself brought him.

*

Despite everything, she tried, according to theories she considered sound, to make herself in love. By moonlight in the garden she used to recite to him all the love poetry she knew, or to sing with a sigh slow melancholy songs. It left her as unmoved as before, neither did it appear to make Charles more loving or more emotional.

Having thus plied the flint for a while without striking a single spark from her heart; being, moreover, as incapable of understanding what she had not experienced as she was of believing in anything that did not present itself in the accepted forms, she had no difficulty in persuading herself that there was nothing very startling now about Charles' passion for her. His ardours had lapsed into a routine, his embraces kept fixed hours; it was just one more habit, a sort of dessert he looked forward to after the monotony of dinner.

*

If matters had fallen out differently, she wondered, might she not have met some other man?

GUSTAVE FLAUBERT, *Madame Bovary*, 1857

The ardent young Dorothea has married Mr Casaubon, an elderly cleric, believing that she could dedicate her life to helping him in his scholarly work. Instead she finds, already on her wedding journey, that her spirit is parched by her life with this dessicated man, and sees through the empty self-importance of his work.

Whatever else remained the same, the light had changed, and you cannot find the pearly dawn at noonday. The fact is unalterable, that a fellow-mortal with whose nature you are acquainted solely through the brief entrances and exits of a few imaginative weeks called courtship, may, when seen in the continuity of married companionship, be disclosed as something better or worse than what you have preconceived, but will certainly not appear altogether the same. And it would be astonishing to find how soon the change is felt if we had no kindred changes to compare with it. To share lodgings with a brilliant dinner-companion, or to see your favourite politician in the Ministry, may bring about changes quite as rapid: in these cases too we begin by knowing little and believing much, and we sometimes end by inverting the quantities.

Still, such comparisons might mislead, for no man was more incapable of flashy make-believe than Mr Casaubon: he was as genuine a character as any ruminant animal, and he had not actively assisted in creating any illusions about himself. How was it that in the weeks since her marriage, Dorothea had not distinctly observed but felt with a stifling depression, that the large vistas and wide fresh air which she had dreamed of finding in her husband's mind were replaced by anterooms and winding passages which seemed to lead nowhither? I suppose it was that in courtship everything is regarded as provisional and preliminary, and the smallest sample of virtue or accomplishment is taken to guarantee delightful stores which the broad leisure of marriage will reveal. But the door-sill of marriage once crossed, expectation is concentrated on the present. Having once embarked on your marital voyage, it is impossible not to be aware that you make no way and that the sea is not within sight—that, in fact, you are exploring an enclosed basin.

*

These characteristics, fixed and unchangeable as bone in Mr Casaubon, might have remained longer unfelt by Dorothea if she had been encouraged to pour forth her girlish and womanly feeling—if he would have held her hands between his and listened with the delight of tenderness and understanding to all the little histories which made up her experience, and would have given her the same sort of intimacy in return, so that the past life of each could be included in their mutual knowledge and affection—or if she could have fed her affection with those childlike caresses which are the bent of every sweet woman, who has begun by showering kisses on the hard pate of her bald doll, creating a happy soul within that woodenness from the wealth of her own love. That was Dorothea's bent. With all her yearning to know what was afar from her and to be widely benignant, she had ardour enough for what was near, to have kissed Mr Casaubon's coat-sleeve, or to have caressed his shoe-latchet, if he would have made any other sign of acceptance than pronouncing her, with his unfailing propriety, to be of a most affectionate and truly feminine nature, indicating at the same time by politely reaching a chair for her that he regarded these manifestations as rather crude and startling. Having made his clerical toilette with due care in the morning, he was prepared only for those amenities of life which were suited to the well-adjusted stiff cravat of the period, and to a mind weighted with unpublished matter.

And by a sad contradiction Dorothea's ideas and resolves seemed like melting ice floating and lost in the warm flood of which they had been but another form. She was humiliated to find herself a mere victim of feeling, as if she could know nothing except through that medium: all her strength was scattered in fits of agitation, of struggle, of despondency, and then again in visions of more complete renunciation, transforming all hard conditions into duty. Poor Dorothea! she was certainly troublesome—to herself chiefly; but this morning for the first time she had been troublesome to Mr Casaubon.

*

Both were shocked at their mutual situation—that each should have betrayed anger towards the other. If they had been at home, settled at Lowick in ordinary life among their neighbours, the clash would have been less embarrassing: but on a wedding journey, the express object of which is to isolate two people on the ground that they are all the world to each other, the sense of disagreement is, to say the least, confounding and stultifying. To have changed your longitude extensively, and placed yourself in a moral solitude in order to have small explosions, to find conversation difficult and to hand a glass of water without looking, can hardly be regarded as satisfactory fulfilment even to the toughest minds. To Dorothea's inexperienced sensitiveness, it seemed like a catastrophe, changing all prospects; and to Mr Casaubon it was a new pain, he never having been on a wedding journey before, or found himself in that close union which was more of a subjection than he had been able to imagine, since this charming young bride not only obliged him to much consideration on her behalf (which he had sedulously given), but turned out to be capable of agitating him cruelly just where he most needed soothing. Instead of getting a soft fence against the cold, shadowy, unapplausive audience of his life, had he only given it a more substantial presence?

*

Mr and Mrs Casaubon, returning from their wedding journey, arrived at Lowick Manor in the middle of January. . . . In the first minutes when Dorothea looked out she felt nothing but the dreary oppression; then came a keen remembrance, and turning away from the window, she walked round the room. The ideas and hopes which were living in her mind when she first saw this room three months before were present now only as memories; she judged them as we judge transient and departed things. All existence seemed to beat with a lower pulse than her own, and her religious faith was a solitary cry, the struggle out of a nightmare in which every object was withering and shrinking away from her. Each remembered thing in the room was disenchanted, was deadened as an unlit transparency . . . The duties of her married life, contemplated as so great beforehand, seemed to be shrinking with the furniture and the white vapour-walled landscape.

*

There was no denying that Dorothea was as virtuous and lovely a young lady as he could have obtained for a wife; but a young lady turned out to be something more troublesome than he had conceived. She nursed him, she read to him, she anticipated his wants, and was solicitous about his feelings; but there had entered into the husband's mind the certainty that she judged him, and that her wifely devotedness was like a penitential expiation of unbelieving thoughts—was accompanied with a power of comparison by which himself and his doings were seen too luminously as a part of things in general. His discontent passed vapour-like through all her gentle loving manifestations, and clung to that inappreciative world which she had only brought nearer to him.

Poor Mr Casaubon! This suffering was the harder to bear because it seemed like a betrayal: the young creature who had worshipped him with perfect trust had quickly turned into the critical wife; and early instances of criticism and resentment had made an impression which no tenderness and submission afterwards could remove. To his suspicious interpretation Dorothea's silence now was a suppressed rebellion; a remark from her which he had not in any way anticipated was an assertion of conscious superiority; her gentle answers had an irritating cautiousness in them; and when she acquiesced it was a self-approved effort of forbearance. The tenacity with which he strove to hide this inward drama made it the more vivid for him; as we hear with the more keenness what we wish others not to hear.

*

Instead of wondering at this result of misery in Mr Casaubon, I think it quite ordinary. Will not a tiny speck very close to our vision blot out the glory of the world, and leave only a margin by which we see the blot? I know no speck so troublesome as self. And who, if Mr Casaubon had chosen to expound his discontents—his suspicions that he was not any longer adored without criticism—could have denied that they were founded on good reasons? On the contrary, there was a strong reason to be added, which he had not himself taken explicitly into account—namely, that he was not unmixedly adorable. He suspected this, however, as he suspected other things, without confessing it, and like the rest of us, felt how soothing it would have been to have a companion who would never find it out.

GEORGE ELIOT, *Middlemarch*, 1871–2

In all matrimonial associations there is, I believe, one constant factor—a desire to deceive the person with whom one lives as to some weak spot in one's character or in one's career. For it is intolerable to live constantly with one human being who perceives one's small meannesses. It is really death to do so—that is why so many marriages turn out unhappily.

FORD MADOX FORD, *The Good Soldier*, 1915

He wanted sympathy. He was a failure, he said. Mrs Ramsay flashed her needles. Mr Ramsay repeated, never taking his eyes from her face, that he was a failure. She blew the words back at him. 'Charles Tansley . . .' she said. But he must have more than that. It was sympathy he wanted, to be assured of his genius, first of all, and then to be taken within the circle of life, warmed and soothed, to have his senses restored to him, his barrenness made fertile, and all the rooms of the house made full of life—the drawing-room; behind the drawing-room the kitchen; above the kitchen the bedrooms; and beyond them the nurseries; they must be furnished, they must be filled with life.

Charles Tansley thought him the greatest metaphysician of the time, she said. But he must have more than that. He must have sympathy. He must be assured that he too lived in the heart of life; was needed; not here only, but all over the world. . . .

He was a failure, he repeated. Well, look then, feel then. Flashing her needles, glancing round about her, out of the window, into the room, at James himself, she assured him, beyond a shadow of a doubt, by her laugh, her poise, her competence (as a nurse carrying a light across a dark room assures a fractious child), that it was real; the house was full; the garden blowing. If he put implicit faith in her, nothing should hurt him; however deep he buried himself or climbed high, not for a second should he find himself without her. . . .

Filled with her words, like a child who drops off satisfied, he said, at last, looking at her with humble gratitude, restored, renewed, that he would take a turn; he would watch the children playing cricket. He went.

Immediately, Mrs Ramsay seemed to fold herself together, one petal closed in another, and the whole fabric fell in exhaustion upon itself, so that she had only strength enough to move her finger, in exquisite abandonment to exhaustion, across the page of Grimm's fairy story, while there throbbed through her, like the pulse in a spring which has expanded to its full width and now gently ceases to beat, the rapture of successful creation.

Every throb of this pulse seemed, as he walked away, to enclose her and her husband, and to give to each that solace which two different notes, one high, one low, struck together, seem to give each other as they combine. Yet, as the resonance died, and she turned to the Fairy Tale again, Mrs Ramsay felt not only exhausted in body (afterwards, not at the time, she always felt this) but also there tinged her physical fatigue some faintly disagreeable sensation with another origin. . . . It came from this: she did not like, even for a second, to feel finer than her husband; and further, could not bear not being entirely sure, when she spoke to him, of the truth of what she said. . . . all this diminished the entire joy, the pure joy, of the two notes sounding together, and let the sound die on her ear now with a dismal flatness.

VIRGINIA WOOLF, *To the Lighthouse*, 1927

For this is one of the miracles of love; it gives—to both, but perhaps especially to the woman—a power of seeing through its own enchantments and yet not being disenchanted.

C. S. LEWIS, *A Grief Observed*, 1961

MEDEA: It was everything to me to think well of one man,
And he, my own husband, has turned out wholly vile.
Of all things which are living and can form a judgment
We women are the most unfortunate creatures.
Firstly, with an excess of wealth it is required
For us to buy a husband and take for our bodies
A master; for not to take one is even worse.
And now the question is serious whether we take
A good or bad one; for there is no easy escape
For a woman, nor can she say no to her marriage.
She arrives among new modes of behavior and manners,
And needs prophetic power, unless she has learned at home,
How best to manage him who shares the bed with her.
And if we work out all this well and carefully,
And the husband lives with us and lightly bears his yoke,
Then life is enviable. If not, I'd rather die.
A man, when he's tired of the company in his home,
Goes out of the house and puts an end to his boredom

And turns to a friend or companion of his own age.
But we are forced to keep our eyes on one alone.
What they say of us is that we have a peaceful time
Living at home, while they do the fighting in war.
How wrong they are! I would very much rather stand
Three times in the front of battle than bear one child.

EURIPIDES, *The Medea*, 431 BC

In the staring gas light, the women, throwing back their shawls from their dishevelled hair revealed faces which, though dissimilar in features, had a similarity of expression common, typical, of all the married women around and about; their badge of marriage, as it were. The vivacity of their virgin days was with their virgin days, gone; a married woman could be distinguished from a single by a glance at her facial expression. Marriage scored on their faces a kind of preoccupied, faded, lack-lustre air as though they were constantly being plagued by some problem. As they were. How to get a shilling, and when obtained, how to make it do the work of two. Though it was not so much a problem as a whole-time occupation to which no salary was attached, not to mention the sideline of risking life to give children birth and being responsible for their upbringing afterwards.

WALTER GREENWOOD, *Love on the Dole*, 1933

AGNES: I thought a little last night, too: while you were seeing everything so clearly here. I lay in the dark, and I . . . revisited—our life, the years and years. There are many things a woman does: she bears the children—if there *is* that blessing. Blessing? Yes, I suppose, even with the sadness. She runs the house, for what that's worth: makes sure there's food, and not just anything, and decent linen; looks well; assumes whatever duties are demanded—if she is in love, or loves; and plans.

TOBIAS [*Mumbled; a little embarrassed*]: I know, I know. . . .

AGNES: And plans. Right to the end of it; expects to be alone one day, abandoned by a heart attack or the cancer, *prepares* for that. And prepares earlier, for the children to become *adult* strangers instead of growing ones, for that loss, and for the body chemistry, the end of what the Bible tells us is our usefulness. The reins we hold! It's a team of twenty horses, and we sit there, and we watch the road and check the leather . . . if our . . . man is so disposed. But there are things we do not do.

TOBIAS [*Slightly edgy challenge*]: Yes?

AGNES: Yes. [*Harder*] We don't decide the route.

TOBIAS: You're copping out . . . as they say.

AGNES: No, indeed.

TOBIAS [*Quiet anger*]: *Yes*, you are!

AGNES [*Quiet warning*]: Don't you yell at me.

TOBIAS: You're copping *out!*

AGNES [*Quiet, calm, and almost smug*]: We follow. We let our . . . men decide the moral issues.

TOBIAS [*Quite angry*]: Never! You've never done that in your life!

AGNES: Always, my darling. Whatever you decide . . . I'll make it work; I'll run it for you so you'll never know there's been a change in anything.

EDWARD ALBEE, *A Delicate Balance*, 1966

Since her children had gone away to school she had wished that she could do her own housework, but that would have meant getting rid of Edith who came in three times a week and relied on her wages for the money to go on holiday to Malta or Ostend with her mother and her husband and the twins. Claudia was glad that Charles was so untidy, because if it hadn't been for his wandering socks, overflowing ashtrays and muddled papers she would have had almost nothing to do. She was bored, and irritated by her own predictability. While she had never thought of herself as an exceptional person, neither had she imagined, as life went by, that she would respond to its vicissitudes with such totally conventional reactions. Now, like someone approaching the end of adolescence, she found herself chafing at the confines of home. Like a liveried servant she began to wish that she could return to her husband all that he had lent her to keep out the cold. She had given her life into his keeping and he had put it away like a garment, neatly cleaned and ironed and never to be worn. In return he had bestowed on her the respectable habiliments of the wife and mother—the apron, the milk-soaked blouse, the blood-stained knickers, the coat made of the skins of animals that was her reward for conjugality, and cold, shining little jewels that she never wore since she had seen other women wearing theirs and had despised them. Those women, in their rings and necklaces, had had a stamped, a franked appearance that did credit neither to them nor to their owners. It had been, to date, her only rebellion and had arisen from taste rather than insight. Stirring within her Claudia now felt the desire to wear what she had never worn—the strange and garish garment of her self. She found she wanted to go about the world in freedom, unrecognisable and subject to no one.

ALICE THOMAS ELLIS, *The Other Side of the Fire*, 1983

MRS BRIDGENORTH: She won't marry, Collins.

COLLINS: Bless you, ma'am, they all say that. You and me said it, I'll lay. I did, anyhow.

MRS BRIDGENORTH: No: marriage came natural to me. I should have thought it did to you too.

COLLINS [*pensive*]: No, ma'am: it didn't come natural. My wife had to break me into it. It came natural to her: she's what you might call a regular old hen. Always wants to have her family within sight of her. Wouldn't go to bed unless she knew they was all safe at home and the door locked, and the lights out. Always wants her luggage in the carriage with her. Always goes and makes the engine driver promise her to be careful. She's a born wife and mother, ma'am. That's why my children all ran away from home.

MRS BRIDGENORTH: Did you ever feel inclined to run away, Collins?

COLLINS: Oh yes ma'am, yes: very often. But when it came to the point I couldn't bear to hurt her feelings. She's a sensitive, affectionate, anxious soul; and she was never brought up to know what freedom is to some people. You see, family life is all the life she knows: she's like a bird born in a cage, that would die if you let it loose in the woods. When I thought how little it was to a man of my easy temper to put up with her, and how deep it would hurt her to think it was because I didn't care for her, I always put off running away 'til next time; and so in the end I never ran away at all. I daresay it was good for me to be took such care of; but it cut me off from all my old friends something dreadful, ma'am: especially the women, ma'am. She never gave them a chance: she didn't indeed. She never understood that married people should take holidays from one another if they are to keep at all fresh. Not that I ever got tired of her, ma'am; but my! how I used to get tired of home life sometimes.

GEORGE BERNARD SHAW, *Getting Married*, 1908

The problem lay buried, unspoken, for many years in the minds of American women. It was a strange stirring, a sense of dissatisfaction, a yearning that women suffered in the middle of the twentieth century in the United States. Each suburban wife struggled with it alone. As she made the beds, shopped for groceries, matched slip-cover material, ate peanut butter sandwiches with her children, chauffeured Cub Scouts and Brownies, lay beside her husband at night—she was afraid to ask even of herself the silent question—'Is this all?'

BETTY FRIEDAN, *The Feminine Mystique*, 1963

NORA: I've been greatly wronged, Torvald. First by my father, and then by you.

HELMER: What! Us two! The two people who loved you more than anybody?

NORA [*shakes her head*]: You two never loved me. You only thought how nice it was to be in love with me.

HELMER: But, Nora, what's this you are saying?

NORA: It's right, you know, Torvald. At home, Daddy used to tell me what he thought, then I thought the same. And if I thought differently, I kept quiet about it, because he wouldn't have liked it. He used to call me his baby doll, and he played with me as I used to play with my dolls. Then I came to live in your house. . . .

HELMER: What way is that to talk about our marriage?

NORA [*imperturbably*]: What I mean is: I passed out of Daddy's hands into yours. You arranged everything to your tastes, and I acquired the same tastes. Or I pretended to . . . I don't really know . . . I think it was a bit of both, sometimes one thing and sometimes the other. When I look back, it seems to me I have been living here like a beggar, from hand to mouth. I lived by doing tricks for you, Torvald. But that's the way you wanted it. You and Daddy did me a great wrong. It's your fault that I've never made anything of my life.

HELMER: Nora, how unreasonable . . . how ungrateful you are! Haven't you been happy here?

NORA: No, never. I thought I was, but I wasn't really.

HELMER: Not . . . not happy!

NORA: No, just gay. And you've always been so kind to me. But our house has never been anything but a play-room. I have been your doll wife, just as at home I was Daddy's doll child. And the children in turn have been my dolls. I thought it was fun when you played with me, just as they thought it was fun when I played with them. That's been our marriage, Torvald.

HELMER: There is some truth in what you say, exaggerated and hysterical though it is. But from now on it will be different. Playtime is over; now comes the time for lessons.

NORA: Whose lessons? Mine or the children's?

HELMER: Both yours and the children's, my dear Nora.

NORA: Ah, Torvald, you are not the man to teach me to be a good wife for you.

HELMER: How can you say that?

NORA: And what sort of qualifications have I to teach the children?

HELMER: Nora!

NORA: Didn't you say yourself, a minute or two ago, that you couldn't trust me with that job.

HELMER: In the heat of the moment! You shouldn't pay any attention to that.

NORA: On the contrary, you were quite right. I'm not up to it. There's another problem needs solving first. I must take steps to educate myself. You are not the man to help me there. That's something I must do on my own. That's why I'm leaving you.

HELMER [*jumps up*]: What did you say?

NORA: If I'm ever to reach any understanding of myself and the things around me, I must learn to stand alone. That's why I can't stay here with you any longer.

HELMER: Nora! Nora!

NORA: I'm leaving here at once. I dare say Kristine will put me up for tonight. . . .

HELMER: You are out of your mind! I won't let you! I forbid you!

NORA: It's no use forbidding me anything now. I'm taking with me my own personal belongings. I don't want anything of yours, either now or later.

HELMER: This is madness!

NORA: Tomorrow I'm going home—to what used to be my home, I mean. It will be easier for me to find something to do there.

HELMER: Oh, you blind, inexperienced . . .

NORA: I must set about *getting* experience, Torvald.

HELMER: And leave your home, your husband and your children? Don't you care what people will say?

NORA: That's no concern of mine. All I know is that this is necessary for *me*.

HELMER: This is outrageous! You are betraying your most sacred duty.

NORA: And what do you consider to be my most sacred duty?

HELMER: Does it take me to tell you that? Isn't it your duty to your husband and your children?

NORA: I have another duty equally sacred.

HELMER: You have not. What duty might *that* be?

NORA: My duty to myself.

HELMER: First and foremost, you are a wife and mother.

NORA: That I don't believe any more. I believe that first and foremost I am an individual, just as much as you are—or at least I'm going to try to be. I know most people agree with you, Torvald, and that's also what it says in books. But I'm not content any more with what most people say, or with what it says in books. I have to think things out for myself, and get things clear.

*

HELMER: You are ill, Nora. You are delirious. I'm half inclined to think you are out of your mind.

NORA: Never have I felt so calm and collected as I do tonight.

HELMER: Calm and collected enough to leave your husband and children?

NORA: Yes.

HELMER: Then only one explanation is possible.

NORA: And that is?

HELMER: You don't love me any more.

NORA: Exactly.

HELMER: Nora! Can you say that!

NORA: I'm desperately sorry, Torvald. Because you have always been so kind to me. But I can't help it. I don't love you any more.

HENRIK IBSEN, *The Doll's House*, 1879

DANIEL AT BREAKFAST

His paper propped against the electric toaster
  (Nicely adjusted to his morning use),
Daniel at breakfast studies world disaster
  And sips his orange juice.

The words dismay him. Headlines shrilly chatter
  Of famine, storm, death, pestilence, decay.
Daniel is gloomy, reaching for the butter.
  He shudders at the way

War stalks the planet still, and men know hunger,
  Go shelterless, betrayed, may perish soon.
The coffee's weak again. In sudden anger
  Daniel throws down his spoon

And broods a moment on the kitchen faucet
  The plumber mended, but has mended ill;
Recalls tomorrow means a dental visit,
  Laments the grocery bill.

Then, having shifted from his human shoulder
  The universal woe, he drains his cup,
Rebukes the weather (surely turning colder),
  Crumples his napkin up
And, kissing his wife abruptly at the door,
Stamps fiercely off to catch the 8:04.

THE 5:32

She said, If tomorrow my world were torn in two,
Blacked out, dissolved, I think I would remember
(As if transfixed in unsurrendering amber)
This hour best of all the hours I knew:
When cars came backing into the shabby station,
Children scuffing the seats, and the women driving
With ribbons around their hair, and the trains arriving,
And the men getting off with tired but practiced motion.

Yes, I would remember my life like this, she said:
Autumn, the platform red with Virginia creeper,
And a man coming toward me, smiling, the evening paper
Under his arm, and his hat pushed back on his head;
And wood smoke lying like haze on the quiet town,
And dinner waiting, and the sun not yet gone down.

PHYLLIS McGINLEY, *Times Three*, 1961

A HOLIDAY

THE WIFE

The house is like a garden,
The children are the flowers,
The gardener should come methinks
And walk among his bowers,
Oh! lock the door on worry,
And shut your cares away,
Not time of year, but love and cheer,
Will make a holiday.

THE HUSBAND

Impossible! You women do not know
The toil it takes to make a business grow.
I cannot join you until very late,
So hurry home, nor let the dinner wait.

THE WIFE

The feast will be like *Hamlet*
Without a Hamlet part:
The home is but a house, dear,
Till you supply the heart.

The Xmas gift I long for
You need not toil to buy;
Oh! give me back one thing I lack—
The love-light in your eye.

THE HUSBAND

Of course I love you, and the children too.
Be sensible, my dear, it is for you
I work so hard to make my business pay.
There, now, run home, enjoy your holiday.

THE WIFE (*turning*)

He does not mean to wound me,
  I know his heart is kind.
Alas! that man can love us
  And be so blind, so blind.
A little time for pleasure,
  A little time for play;
A word to prove the life of love
  And frighten Care away!
Tho' poor my lot in some small cot
  That were a holiday.

THE HUSBAND (*musing*)

She has not meant to wound me, nor to vex—
Zounds! but 'tis difficult to please the sex.
I've housed and gowned her like a very queen
Yet there she goes, with discontented mien.
I gave her diamonds only yesterday:
Some women are like that, do what you may.

ELLA WHEELER WILCOX *(1850–1919)*

All seemed well. The days passed, and the weeks, and the months, more swiftly than in childhood, and she felt no trepidation, except for certain moments in the depth of the night when, as she and her new husband lay drowsily clutching each other for reassurance, anticipating the dawn, the day, and another night which might prove them both immortal, Mrs Bridge found herself wide awake. During these moments, resting in her husband's arms, she would stare at the ceiling, or at his face, which sleep robbed of strength, with an uneasy expression, as though she saw or heard some intimation of the great years ahead.

She was not certain what she wanted from life, or what to expect from it, for she had seen so little of it, but she was sure that in some way—because she willed it to be so—her wants and her expectations were the same.

For a while after their marriage she was in such demand that it was not unpleasant when he fell asleep. Presently, however, he began sleeping all night, and it was then she awoke more frequently, and looked into the darkness, wondering about the nature of men, doubtful of the future, until at last there came a night when she shook her husband awake and spoke of her own desire. Affably he placed one of his long white arms around her waist; she turned to him then, contentedly, expectantly, and secure. However nothing else occurred, and in a few minutes he had gone back to sleep.

This was the night Mrs Bridge concluded that while marriage might be an equitable affair, love itself was not.

*

Often he thought: My life did not begin until I knew her.

She would like to hear this, he was sure, but he did not know how to tell her. In the extremity of passion he cried out in a frantic voice: 'I love you!' yet even these words were unsatisfactory. He wished for something else to say. He needed to let her know how deeply he felt her presence while they were lying together during the night, as well as each morning when they awoke and in the evening when he came home. However, he could think of nothing appropriate.

So the years passed, they had three children and accustomed themselves to a life together, and eventually Mr Bridge decided that his wife should expect nothing more of him. After all, he was an attorney rather than a poet; he could never pretend to be what he was not.

*

As time went on she felt an increasing need for reassurance. Her husband had never been a demonstrative man, not even when they were first married; consequently she did not expect too much from him. Yet there were moments when she was overwhelmed by a terrifying, inarticulate need. One evening as she and he were finishing supper together, alone, the children having gone out, she inquired rather sharply if he loved her. She was surprised by her own bluntness and by the almost shrewish tone of her voice, because that was not the way she actually felt. She saw him gazing at her in astonishment; his expression said very clearly: Why on earth do you think I'm here if I don't love you? Why aren't I somewhere else? What in the world has got into you?

Mrs Bridge smiled across the floral centerpiece—and it occurred to her that these flowers she had so carefully arranged on the table were what separated her from her husband—and said, a little wretchedly, 'I know it's silly, but it's been such a long time since you told me.'

Mr Bridge grunted and finished his coffee. She knew it was not that he was annoyed, only that he was incapable of the kind of declaration she needed. It was so little, and yet so much. While they sat across from each other, neither knowing quite what to do next, she became embarrassed; and in her embarrassment she moved her feet and she inadvertently stepped on the buzzer, concealed beneath the carpet, that connected with the kitchen, with the result that Harriet soon appeared in the doorway to see what it was that Mrs Bridge desired.

EVAN S. CONNELL, *Mrs Bridge and Mr Bridge*, 1959

TWO IN THE CAMPAGNA

I

I wonder do you feel to-day
As I have felt since, hand in hand,
We sat down on the grass, to stray
In spirit better through the land,
This morn of Rome and May?

II

For me, I touched a thought, I know,
Has tantalized me many times,
(Like turns of thread the spiders throw
Mocking across our path) for rhymes
To catch at and let go.

III

Help me to hold it! First it left
The yellowing fennel, run to seed
There, branching from the brickwork's cleft,
Some old tomb's ruin: yonder weed
Took up the floating weft,

IV

Where one small orange cup amassed
  Five beetles,—blind and green they grope
Among the honey-meal: and last,
  Everywhere on the grassy slope
I traced it. Hold it fast!

V

The champaign with its endless fleece
  Of feathery grasses everywhere!
Silence and passion, joy and peace,
  An everlasting wash of air—
Rome's ghost since her decease.

VI

Such life here, through such lengths of hours,
  Such miracles performed in play,
Such primal naked forms of flowers,
  Such letting nature have her way
While heaven looks from its towers!

VII

How say you? Let us, O my dove,
  Let us be unashamed of soul,
As earth lies bare to heaven above!
  How is it under our control
To love or not to love?

VIII

I would that you were all to me,
  You that are just so much, no more.
Nor yours nor mine, nor slave nor free!
  Where does the fault lie? What the core
O' the wound, since wound must be?

IX

I would I could adopt your will,
  See with your eyes, and set my heart
Beating by yours, and drink my fill
  At your soul's springs,—your part my part
In life, for good and ill.

X

No. I yearn upward, touch you close,
  Then stand away. I kiss your cheek,
Catch your soul's warmth,—I pluck the rose
  And love it more than tongue can speak—
Then the good minute goes.

XI

Already how am I so far
  Out of that minute? Must I go
Still like the thistle-ball, no bar,
  Onward, whenever light winds blow,
Fixed by no friendly star?

XII

Just when I seemed about to learn!
  Where is the thread now? Off again!
The old trick! Only I discern—
  Infinite passion, and the pain
Of finite hearts that yearn.

ROBERT BROWNING, 1855

Love one another, but make not a bond of love:
Let it rather be a moving sea between the shores of your souls.
Fill each other's cup but drink not from one cup.
Give one another of your bread but eat not from the same loaf.
Sing and dance together and be joyous, but let each one of you be alone,
Even as the strings of a lute are alone though they quiver with the same music.

Give your hearts, but not into each other's keeping.
For only the hand of Life can contain your hearts.
And stand together yet not too near together:
For the pillars of the temple stand apart,
And the oak tree and the cypress grow not in each other's shadow.

KAHLIL GIBRAN, *The Prophet*, 1923

5

With my Body I thee Worship

To the Nuptial Bowre
I led her blushing like the Morn: all Heav'n,
And happie Constellations on that houre
Shed their selectest influence; the Earth
Gave sign of gratulation, and each Hill;
Joyous the Birds; fresh Gales and gentle Aires
Whisper'd it to the Woods, and from thir wings
Flung Rose, flung Odours from the spicie Shrub,
Disporting, till the amorous Bird of Night
Sung Spousal, and bid haste the Eevning Starr
On his Hill top, to light the bridal Lamp.

(John Milton, *Paradise Lost*)

AFTER such rapturous beginnings, Adam and Eve became ashamed of their nakedness, and most of mankind has felt ambivalent about sexuality ever since. In the Judaeo-Christian tradition, sex and shame are closely related: the Jewish Code of Law forbade intercourse by daylight or by the light of a candle, or indeed 'to glance at that place'.

Havelock Ellis wrote: 'I am ever more astonished at the rarity of erotic personality and the ignorance of the art of love.' As A. S. Neill comments in *Summerhill*: 'Considering the infant training that the majority of people have had, it is a matter for astonishment that there should be any happy marriages at all. If sex is dirty in the nursery, it cannot be very clean in the wedding bed', and though infant training has greatly changed since Neill's day, shame can still overshadow the enjoyment of sexuality. Just how painful the consequence of such shame, often born of ignorance, can be, and how destructive to a marriage, is illustrated in Charlotte Mew's poem.

There have been successful marriages in which sex played little or no part—Shaw's marriage is a notable example—but the sexual rapport is a sensitive barometer of a relationship as a whole. Strange, therefore, that so little has been written about sex, and particularly enjoyable sex, within marriage. Poets have celebrated marital sexual love, but few works of fiction depict a happy sexual relationship between man and wife (perhaps because good marital sex does little to advance a plot) and biographers have until recently shrunk from revealing matters of such intimacy.

Sex may be used as a threat, as a trap, as a weapon, or a bargaining counter in a marital power struggle—Aristophanes' Lysistrata has had many followers down the ages—as a means of avoiding confrontation or of mending a quarrel. But at its best, sexual love is a source of mutual affirmation, pleasure, and comfort, and the most joyful expression of married love.

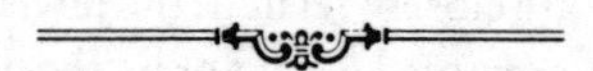

Let thy fountain be blessed: and rejoice with the wife of thy youth.

Let her be as the loving hind and pleasant roe; let her breasts satisfy thee at all times; and be thou ravished always with her love.

Proverbs 5: 18–19

Abstinence sows sand all over
The ruddy limbs and flaming hair,
But Desire Gratified
Plants fruits of life and beauty there.

WILLIAM BLAKE, 1787

MY LORD SUMMONS ME

My lord is all aglow. In his left hand he holds the reed-pipe, with his right he summons me to make free with him. Oh, the joy!

My lord is care-free. In his left hand he holds the dancing plumes, with his right he summons me to sport with him. Oh, the joy!

The Book of Songs (800–600 BC), *trans.* ARTHUR WALEY, 1937

ON THE MARRIAGE OF A VIRGIN

Waking alone in a multitude of loves when morning's light
Surprised in the opening of her nightlong eyes
His golden yesterday asleep upon the iris
And this day's sun leapt up the sky out of her thighs
Was miraculous virginity old as loaves and fishes,
Though the moment of a miracle is unending lightning
And the shipyards of Galilee's footprints hide a navy of doves.
No longer will the vibrations of the sun desire on
Her deepsea pillow where once she married alone,
Her heart all ears and eyes, lips catching the avalanche
Of the golden ghost who ringed with his streams her mercury
bone,

Who under the lids of her windows hoisted his golden luggage,
For a man sleeps where fire leapt down and she learns through
 his arm
That other sun, the jealous coursing of the unrivalled blood.

DYLAN THOMAS, 1941

THE QUESTION ANSWER'D

What is it men in women do require?
The lineaments of Gratified Desire.
What is it women do in men require?
The lineaments of Gratified Desire.

WILLIAM BLAKE, 1787

IN PRAISE OF MARRIAGE

Marriage is a sweet state,
I can affirm it by my own experience,
In very truth, I who have a good and wise husband
Whom God helped me to find.
I give thanks to him who will save him for me,
For I can truly feel his great goodness
And for sure the sweet man loves me well.

Throughout that first night in our home,
I could well feel his great goodness,
For he did me no excess
That could hurt me.
But, before it was time to get up,
He kissed me 100 times, this I affirm,
Without exacting further outrage,
And yet for sure the sweet man loves me well.

He used to say to me in his soft language:
'God brought you to me,
Sweet lover, and I think he raised me
To be of use to you.'

And then he did not cease to dream
All night, his conduct was so perfect,
Without seeking other excesses.
And yet for sure the sweet man loves me well.

O Princes, yet he drives me mad
When he tells me he is all mine;
He will destroy me with his gentle ways,
And yet for sure the sweet man loves me well.

CHRISTINE DE PISAN (1363–1430)

An undated letter, written in the late 1680s, by the Revd William Nevar, to his former pupils Ashe and William Windham:

Sir

I date this Letter from the happiest day of my life, a Levitical Conjurer transformed me this morning from an Insipid, Unrelishing Batchelour into a Loving Passionate Husband, but in the midst of all the raptures of approaching Joys, some of my thoughts must fly to Felbrigg, and tho I am calld away 17 times in a minute to new exquisite dainties, yet I cannot resist the inticing temptation of conversing with you, and acquainting you, with tears in my Eyes, that I am going to lose my Maidenhead, but you'll think perhaps of the old Saying, that some for Joy do cry, and some for Sorrow sing. Colonel Finch, who honours us with his merry company, tells me of dismall dangers I am to run before the next Sun shines upon me, but the Spouse of my bosom being of a meek, forgiving temper, I hope she will be mercifull, and not suffer a young beginner to dye in the Experiment. I commend myself to your best prayers in this dreadfull Juncture, and wishing you speedily such a happy night, as I have now in prospect

I remain
Your most humble and
most obedient Servant
W. Nevar

These letters, exchanged between Charles Kingsley (preacher and canon at Westminster Abbey as well as writer) and his wife Fanny, both before and after their marriage, show the strong erotic bond between them, stimulated perhaps by the initial self-denial, bordering on brinkmanship. Their passionate love continued through many years of married life.

Charles to Fanny: October 1843

Darling, one resolution I made in my sorrow, that I would ask a boon of you and I wish to show you and my God that I have gained purity and self-control, that intense though my love is for your body, I do not love it but as an expression of your soul. And therefore, when we are married, will you consent to remain for the first month in my arms a virgin bride, a sister only? . . . Will not these thoughts give us more perfect delight when we lie naked in each other's arms, clasped together toying with each other's limbs, buried in each other's bodies, struggling, panting, dying for a moment. Shall we not feel then, even then, that there is more in store for us, that those thrilling writhings are but dim shadows of a union which shall be perfect?

20 October 1843

What can I do but write to my naughty baby who does not love me at all and who of course has forgotten me by this time? But I have not forgotten her, for my hands are perfumed with her delicious limbs, and I cannot wash off the scent, and every moment the thought comes across me of those mysterious recesses of beauty where my hands have been wandering, and my heart sinks with a sweet faintness and my blood tingles through every limb for a moment and then all is still again in calm joy and thankfulness to our loving God. Tomorrow I fast, not entirely, but as much as I can without tiring myself. Only to acquire self-control and to keep under the happy body, to which God has permitted of late such exceeding liberty and bliss.

Fanny to Charles: 30 December 1843

After dinner I shall perhaps feel worn out so I shall just lie on your bosom and say nothing but feel a great deal, and you will be very loving and call me your poor child. And then you will perhaps show me your *Life of St Elizabeth*, your wedding gift. And then after tea we will go up to rest! We will undress and bathe and then you will come to my room, and we will kiss and love very much and read psalms aloud together, and then we will kneel down and pray in our night dresses. Oh! What solemn bliss! How hallowing! And then you will take me up in your arms, will you not? And lay me down in bed. And then you will extinguish our light and *come to me*! How I will open my arms to you and then sink into yours! And you will kiss me and clasp me and we will both praise God alone in the dark night with His eye shining down upon us and His love enclosing us. After a time we shall sleep!

And yet I fear you will yearn so for *fuller* communion that you will not be so happy as me. And I too perhaps shall yearn, frightened as I am! But every yearning will remind me of our self-denial, your sorrow for sin, your strength of repentance. And I shall glory in my yearning, *please God*!

Seven years after their marriage Charles wrote to Fanny:

24 July 1857

Oh that I were with you, or rather you with me here. The beds are so small that we should be forced to lie inside each other, and the weather is so hot that you might walk about naked all day, as well as night—*cela va sans dire*! Oh, those naked nights at Chelsea! When will they come again? I kiss both locks of hair every time I open my desk—but the little curly one seems to bring me nearer to you.

Try to prevent your son-in-law from brutalizing your daughter on their wedding night, for many physical weaknesses and painful childbirths among delicate women stem from this cause alone. Men do not sufficiently understand that their pleasure is our martyrdom. So tell him to restrain his pleasure and to wait until he has little by little brought his wife to understanding and response. Nothing is more horrible than the terror, the sufferings, and the revulsion of a poor girl, ignorant of the facts of life, who finds herself raped by a brute. As far as possible we bring them up as saints, and then we hand them over as if they were fillies. If your son-in-law is an intelligent man, and if he truly loves your daughter, he will understand what must be done, and he will not resent your talking it over with him the day before.

GEORGE SAND, from a letter to her half-brother, Hippolyte Chatiron, 1843

A DAUGHTER'S DIFFICULTIES AS A WIFE: MRS REUBEN CHANDLER TO HER MOTHER IN NEW ORLEANS

September 3, 1840 Cincinnati, Ohio

Now that I've been married for almost four weeks, Mama,
I'd better drop you and Papa dear a line.
I guess I'm fine.

Ruby has promised to take me to the Lexington
buggy races Tuesday, if the weather cools.
So far we've not been out much.

Just stayed here stifling in hot Cincinnati.
Clothes almost melt me, Mama, so I've not got out
my lovely red velvet-and-silk pelisse yet,

or that sweet little lambskin coat with the fur hood.
The sheets look elegant!
I adore the pink monogram on the turnover

with exactly the same pattern on the pillowcases!
Darlings!
How I wish you could breeze in and admire them!

*

I'm writing this in bed because
my head thumps and drums every time I move
and I'm so dog tired!

The only time I sleep is in the morning
when Reuben has left for the office.
Which brings up a *delicate* subject, Mama.

I've been thinking and thinking,
wondering whether I'll *ever* succeed in being
the tender, devoted little wife you wanted me to be.

Because . . . oh, Mama,
  why didn't you tell me or warn me before I was married
    that a wife is expected to do it *every night*!

But how could we have guessed?
  Ruby came courting so cool and fine and polite,
    while beneath that gentlemanly, educated exterior . . .

well! I don't like to worry you, Mama.
  You know what men are like!
    I remember you said once the dears couldn't help it.

I try to be brave.
  But if you *did* have a chance to speak to Papa,
    mightn't you ask him to slip a word,

sort of man to man to Reuben . . .
  about how delicate I am,
    and how sick I am every month,

not one of those cows
  who can be used and used?
    Someone's at the door.

I forgot,
  I asked Fanny Daniels to come up this morning
    to help fix a trim for my hat.

I'll have to hustle!
  Give all my love to dear Spooky and Cookie.
    How I miss them, the doggy darlings!

Oceans of hugs and kisses for you, too,
  and for precious Papa,

    From your suffering and loving daughter,

      Marianne

ANNE STEVENSON, 1974

THE FARMER'S BRIDE

Three summers since I chose a maid,
Too young maybe—but more's to do
At harvest-time than bide and woo.
When us was wed she turned afraid
Of love and me and all things human;
Like the shut of a winter's day.
Her smile went out and 'twasn't a woman—
More like a little frightened fay.
One night, in the Fall, she runned away.

'Out 'mong the sheep, her be,' they said,
'Should properly have been abed;
But sure enough she wasn't there
Lying awake with her wide brown stare.
So over seven-acre field and up-along across the down
We chased her, flying like a hare
Before our lanterns. To Church-Town
All in a shiver and a scare
We caught her, fetched her home at last
And turned the key upon her, fast.

She does the work about the house
As well as most, but like a mouse:
Happy enough to chat and play
With birds and rabbits and such as they,
So long as men-folk keep away.

'Not near, not near,' her eyes beseech
When one of us comes within reach.
The women say that beasts in stall
Look round like children at her call.
I've hardly heard her speak at all.

Shy as a leveret, swift as he,
Straight and slight as a young larch tree,
Sweet as the first wild violets, she,
To her wild self. But what to me?

The short days shorten and the oaks are brown,
  The blue smoke rises to the low grey sky,
One leaf in the still air falls slowly down,
  A magpie's spotted feathers lie
On the black earth spread white with rime,
The berries redden up to Christmas time.
  What's Christmas time without there be
  Some other in the house than we!

  She sleeps up in the attic there
  Alone, poor maid. 'Tis but a stair
Betwixt us. Oh! my God! the down,
  The soft young down of her, the brown,
The brown of her—her eyes, her hair, her hair!

CHARLOTTE MEW, 1915

THE WINE-SELLER'S WIFE

There was a wine-seller's shop, as you went down to the river in the city of the Anti-popes. . . .

They called the wine-seller Paradou. He was built more like a bullock than a man, huge in bone and brawn, high in colour, and with a hand like a baby for size. Marie-Madeleine was the name of his wife; she was of Marseilles, a city of entrancing women, nor was any fairer than herself. She was tall, being almost of a height with Paradou; full-girdled, point-device in every form, and with an exquisite delicacy in the face; her nose and nostrils a delight to look at from the fineness of the sculpture, her eyes inclined a hair's-breadth inward, her colour between dark and fair, and laid on even like a flower's. A faint rose dwelt in it, as though she had been found unawares bathing, and had blushed from head to foot. She was of a grave countenance, rarely smiling; yet it seemed to be written upon every part of her that she rejoiced in life. Her husband loved the heels of her feet and the knuckles of her fingers; he loved her like a glutton and a brute; his love hung about her like an atmosphere; one that came by chance into the wine-shop was aware of that passion; and it might be said that by the strength of it the woman had been drugged or spellbound. She knew not if she loved or loathed him; he was always in her eyes like something monstrous—monstrous in his love, monstrous in his person, horrific but imposing in his violence; and her sentiment swung back and forward from desire to sickness. But the mean, where it dwelt chiefly, was an apathetic fascination, partly of horror; as of Europa in mid ocean with her bull.

ROBERT LOUIS STEVENSON, *Lay Morals*, 1918

Wives are always allowed their humour, yet it is only in exchange for titillation and pleasure, which indeed are but other names for Folly; as none can deny, who consider how a man must hug, and dandle, and kittle, and play a hundred little tricks with his bed-fellow when he is disposed to make that use of her that nature designed her for. Well then, you see whence that greatest pleasure (to which modesty scarce allows a name) springs and proceeds.

ERASMUS, *The Praise of Folly*, 1511

THE AUTHOR TO HIS WIFE, OF A WOMAN'S ELOQUENCE

My Mall, I mark that when you mean to prove me
To buy a velvet gown, or some rich border,
Thou call'st me good sweet heart, thou swear'st to love me,
Thy locks, thy lips, thy looks, speak all in order,
Thou think'st, and right thou think'st, that these do move me,
That all these severally thy suit do further:
　But shall I tell thee what most thy suit advances?
　Thy fair smooth words? no, no, thy fair smooth haunches.

SIR JOHN HARINGTON (*c.*1561–1612)

'WIFE, THERE ARE SOME POINTS'
(Epigram 104, Book XI)

Wife, there are some points on which we differ from each other:
Either change your ways to mine or go back to your mother.
She, I grant, has brought you up to be a model daughter—
Bed at sunset, up at daybreak, drinking only water—
I, however, am no antique Roman, stiff and stern:
To me the night's for making love and drinking wine in turn.

You wear a nightgown, robe and girdle, even in hot weather:
I like sleeping with a woman in the altogether.
I want kisses long drawn out, lips parted, tongue-tips meeting:
Yours could just as well be your grandmother's morning
greeting.
You think we should be in total darkness when we're screwing:
I prefer a light; I like to see what I am doing.
Women should co-operate, caressing, squirming, squealing;
You lie back and think of Rome, or contemplate the ceiling.
Hector's house-slaves masturbated as, through half-shut doors,
They watched Andromache, astraddle, ride her hubby-horse.
That paragon of chastity, Penelope, would keep
Her hand there after Ulysses was snoring in his sleep.
You won't pretend to be a boy; such tricks, you say, don't
suit us:
Julia lay prone for Pompey, Portia did for Brutus;
Even Jupiter, before he had his Ganymede,
Persuaded Juno to turn round to satisfy his need.
By all means be Lucrece and sit like Virtue on a dais
All day long, provided that in bed at night you're Lais.

MARTIAL, *c.* AD 98

She loved it when Brian blew into her ear, gently bit the base of her thumb, or stroked her breasts in circles. She sighed and smiled and stretched like a cat when he licked a slow line down the length of her spine, and further. 'Oh love, love,' she murmured. 'Oh bliss.' If only he had been satisfied to stop there, he could feel her thinking. But no, he always had to bring out, or up, what she called 'that thing'. 'Don't put that thing in yet please, darling; I'm not ready.'

'My cock, my prick, my penis for God's sake,' he had shouted at her once. 'Can't you call it by its right name?' No, she couldn't. She didn't like any of those words; she never thought them in her mind and she couldn't say them. She knew words for the other difficult parts of the body: 'behind' for ass and 'stomach' for belly, but there was no word for That Thing. Or occasionally, when Erica was really hurt or annoyed with him, Your Thing. Ordinarily, out of good manners, she overlooked the fact of his connection with the Thing, and when possible its very existence. She avoided looking at it directly, and never touched it unless she was specifically requested to do so. It was as if Brian were a neighbour who owned a particularly ugly dog. '*The* dog is scratching at the door,' you might say to him politely, not wishing to underline their relationship—but, in anger, 'Your dog bit me.'

Two years passed in this way. Then Erica became pregnant. Her obstetrician, a cautious, prissy man, advised that she 'avoid intercourse' from the sixth month on, and for two months after the birth—in effect, a five-month abstinence. . . .

But the woman who was restored to him when Jeffo was eight weeks old was worth waiting for. Whether the cause was physiological or psychological, Erica had matured sexually. She retained her verbal modesty, but now she spoke of Brian's organ gently and affectionately as 'it', and in moments of enthusiasm as 'he'. For fifteen years (with five more months off for Matilda) they had made each other happy.

Now it is as if the bad, half-forgotten early period of their marriage had returned. In bed Erica is compliant; but He is called Thing again, and under the soft rhythm of her pleasure Brian thinks he can hear a counter-beat: the heavy creaking and thumping of a deadly struggle between his will to enter and her will to delay the invasion as long as possible, so that the occupation might be as short as possible. His main weapons in this battle are force and persuasion; Erica's fuss and delay. She can't get into bed at night now until she is sure, absolutely sure, that the doors are locked, the gas turned off, the thermostat down, the cat shut in the pantry with a full box of Kitty-Litter, and the children sleeping soundly and warmly covered. Then it takes her up to five minutes to find and insert her diaphragm (she refuses to go on the pill because of blood-clots), and longer to get out of her nightgown than it takes Wendy to undress completely. And these are only the preliminary manoeuvres.

A real victory for Erica took place on the few occasions when she was able to hold back the invading troops for so long that, fatigued and impatient, they discharged all their artillery at the frontier. But real victory for either side is rare. Usually, rather than face Erica's wounded body next day ('I'm still a little sore down there,' she would say, placing a cushion on her kitchen chair) he held back for a while. And she, rather than face his wounded spirit, finally gave way; but she gave way condescendingly, with a characteristic *noblesse oblige*.

ALISON LURIE, *The War Between the Tates*, 1974

'Tell me to what conclusion or in aid
Of what were generative organs made?
And for what profit were those creatures wrought?
Trust me, they cannot have been made for naught.
Gloze as you will and plead the explanation
That they were only made for the purgation
Of urine, little things of no avail
Except to know a female from a male,
And nothing else. Did somebody say no?
Experience knows well it isn't so.
The learned may rebuke me, or be loth
To think it so, but they were made for both,
That is to say both use and pleasure in
Engendering, except in case of sin.
Why else the proverb written down and set
In books: 'A man must yield his wife her debt'?
What means of paying her can he invent
Unless he use his silly instrument?
It follows they were fashioned at creation
Both to purge urine and for propagation.
'But I'm not saying everyone is bound
Who has such harness as you heard me expound
To go and use it breeding; that would be
To show too little care for chastity.
Christ was a virgin, fashioned as a man,
And many of his saints since time began
Were ever perfect in their chastity.
I'll have no quarrel with virginity.
Let them be pure wheat loaves of maidenhead
And let us wives be known for barley-bread;
Yet Mark can tell that barley-bread sufficed
To freshen many at the hand of Christ.
In that estate to which God summoned me
I'll persevere; I'm not pernickety.
In wifehood I will use my instrument
As freely as my Maker me it sent.
If I turn difficult, God give me sorrow!
My husband, he shall have it eve and morrow
Whenever he likes to come and pay his debt,
I won't prevent him! I'll have a husband yet

Who shall be both my debtor and my slave
And bear his tribulation to the grave
Upon his flesh, as long as I'm his wife.
For mine shall be the power all his life
Over his proper body, and not he,
Thus the Apostle Paul has told it me,
And bade our husbands they should love us well;
There's a command on which I like to dwell . . .'

GEOFFREY CHAUCER, *The Wife of Bath's Tale: Prologue*, c.1390

Come, madam, come, all rest my powers defy;
Until I labour, I in labour lie.
The foe ofttimes, having the foe in sight,
Is tired with standing, though he never fight.
Off with that girdle, like heaven's zone glittering,
But a far fairer world encompassing.
Unpin that spangled breast-plate, which you wear,
That th'eyes of busy fools may be stopp'd there.
Unlace yourself, for that harmonious chime
Tells me from you that now it is bed-time.
Off with that happy busk, which I envy,
That still can be, and still can stand so nigh.
Your gown going off such beauteous state reveals,
As when from flowery meads th'hill's shadow steals.
Off with your wiry coronet, and show
The hairy diadems which on you do grow.
Off with your hose and shoes; then softly tread
In this love's hallow'd temple, this soft bed.
In such white robes heaven's angels used to be
Revealed to men; thou, angel, bring'st with thee
A heaven-like Mahomet's paradise; and though
Ill spirits walk in white, we easily know
By this these angels from an evil sprite;
Those set our hairs, but these our flesh upright.
  Licence my roving hands, and let them go
Before, behind, between, above, below.
Oh, my America, my Newfoundland,
My kingdom, safest when with one man mann'd,
My mine of precious stones, my empery;

How am I blest in thus discovering thee!
To enter in these bonds, is to be free;
Then, where my hand is set, my seal shall be.
  Full nakedness! All joys are due to thee;
As souls unbodied, bodies unclothed must be
To taste whole joys. Gems which you women use
Are like Atlanta's ball cast in men's views;
That, when a fool's eye lighteth on a gem,
His earthly soul might court that, not them.
Like pictures, or like books' gay coverings made
For laymen, are all women thus array'd.
Themselves are only mystic books, which we
—Whom their imputed grace will dignify—
Must see reveal'd. Then, since that I may know,
As liberally as to thy midwife show
Thyself; cast all, yea, this white linen hence;
There is no penance due to innocence:
  To teach thee, I am naked first; why than,
What needst thou have more covering than a man?

JOHN DONNE, 'Elegy XIX', pub. 1633

SONG OF A MAN WHO IS LOVED

Between her breasts is my home, between her breasts.
Three sides set on me space and fear, but the fourth side rests,
Warm in a city of strength, between her breasts.

All day long I am busy and happy at my work
I need not glance over my shoulder in fear of the terrors that
  lurk
Behind. I am fortified, I am glad at my work.

I need not look after my soul; beguile my fear
With prayer, I need only come home each night to find the dear
Door on the latch, and shut myself in, shut out fear.

I need only come home each night and lay
My face between her breasts;
And what of good I have given the day, my peace attests.

And what I have failed in, what I have wronged
Comes up unnamed from her body and surely
Silent tongued I am ashamed.

And I hope to spend eternity
With my face down-buried between her breasts
And my still heart full of security
And my still hands full of her breasts.

D. H. LAWRENCE, 1928

In our bed in the villa by the Mediterranean my husband slid from my body and said, 'How I hate all Wagner. *Tristan and Isolde* is nothing like it, is it? It is so sharp and clear, and the *Tristan* music is like two fat people eating thick soup. Drinking thick soup,' he corrected himself pedantically, and yawned, and nuzzled his face against my shoulder, and was asleep. I ran my arm down his straight back. When I had thought of his face in the train to Reading, it had seemed to me more right than nature could make it, it was as if a master craftsman had worked on it. His body was like that too. I enjoyed everything about being married, though I could not have endured it with any other husband but Oliver. I was amazed at lovemaking. It was so strange to come, when I was nearly middle-aged, on the knowledge that there was another state of being than any I had known, and that it was the state normal for humanity, that I was a minority who did not know it. It was as if I had learned that there was a sixth continent, which nearly everybody but me and a few others had visited and in which, now I had come to it, I felt like a native, or as if there was another art as well as music and painting and literature, which was not only preached, but actually practised, by nearly everybody, though they were silent about their accomplishment. It was fantastic that nobody should speak of what pervaded life and determined it, yet it was inevitable, for language could not describe it. I looked across Oliver at the window, which we had opened after we had put out the light and there was no fear of attracting mosquitoes. There was the sea, glittering with moonlight, the dark mountains above it, then the sky dusted with other earths, which looking at us might not know that our globe was swathed in this secret web of nakedness that kept it from being naked of people, chilly with lack of love and life, a barren top spinning to no purpose. Their architecture would be as fantastic but would not be the same, because there were not two of anything alike, every person was different, every work of art was different, every act of love was different, every world was different. It was a pity we did not know the end to which this wealth was to be put, but surely if this plenitude existed, and not the nothingness which somehow seemed to be more natural, more what one would have expected (though it is the one state of which the universe had and could have no experience), we might conclude that all would be well. I could believe that this precious intricate creature I held in my arms, who made love and wrote music, would never be destroyed.

REBECCA WEST, *Cousin Rosamund*, 1957

May 25, 1932

Darling Mistress Fatbum,

. . . Your three letters mailed Monday arrived—and were, as usual, the big, desirable event of the day. Why do you say you hesitated to phone Sunday night because you knew I hated it? Not with you at the other end, I don't! If you knew how thrilled and delighted I was! Haven't you discovered that I love you yet, Foolish One? . . .

Soul & body & all of me lives in you! Mistress, I desire you, you are my passion, and my life-drunkenness, and my ecstasy, and the wine of joy to me! Wife, you are my love, and my happiness, and the word behind my word, and the half of my heart! Mother, you are my lost way refound, my end and my beginning, the hand I reach out for in my lonely night, from my ghost-haunted inner dark, and on your soft breasts there is a peace for me that is beyond death! Daughter, you are my secret, shy, shrinking one, my pure and unsoiled one, whom the world has wounded and who still looks with bewildered, hurt eyes at the world and at her wound and cannot understand; but I cuddle you in my arms, and you are blood of my blood and flesh of my flesh, and over your comforted, sleeping head against my heart swear my oath at life, that my life shall be between you and all wounds and hurt hereafter.

EUGENE O'NEILL, letter to his wife Carlotta

Love is something far more than desire for sexual intercourse; it is the principal means of escape from the loneliness which affects most men and women throughout the greater part of their lives. There is a deep-seated fear, in most people, of the cold world and the possible cruelty of the herd; there is a longing for affection, which is often concealed by roughness, boorishness or a bullying manner in men, and by nagging and scolding in women. Passionate mutual love while it lasts puts an end to this feeling; it breaks down the hard walls of the ego, producing a new being composed of two in one. Nature did not construct human beings to stand alone, since they cannot fulfil her biological purpose except with the help of another; and civilized people cannot fully satisfy their sexual instinct without love. The instinct is not completely satisfied unless a man's whole being, mental quite as much as physical, enters into the relation. Those who have never known the deep intimacy and the intense companionship of mutual love have missed the best thing that life has to give; . . . if they miss this experience, men and women cannot attain their full stature, and cannot feel towards the rest of the world that kind of generous warmth without which their social activities are pretty sure to be harmful.

BERTRAND RUSSELL (1872–1970), 'Love, An Escape from Loneliness'

In a fulfilling sexual act, the two opposite genders, at their most different and separate, simultaneously become one totality, merging with one another in an experience going beyond the capacity of either. Each is most centered in, and aware of, himself or herself, yet also wholly open and responsive to the other. Each temporarily loses his boundary, surrenders to a greater unity. Both are as spontaneous as they could ever be, yet this spontaneity is possible because of a fundamental self-discipline, an ability to deny oneself, to wait, to adapt and adjust to the other as in the unfolding of a dance. It is non-manipulative, non-controlling; the self is offered freely, from generosity and trust, and since there is no demand the return comes equally freely and fully, each emotionally responding and keeping time with the other. Each gives most generously, yet takes most uninhibitedly too, without hedging or bargaining. For a moment, both partners are fully in the present, letting the process unfold in its own form and pace, no longer lost in the concerns of the future or memories of the past. And the climax, when it 'comes' in its own time, is productive, creative, sometimes through the beginning of a separate new life but always in a renewal of the separate lives of the partners and the joint life of their relationship, so that the wild, tender, simple act of affirmation is never tired of, never loses its fullness or the refreshing quality of a draught of spring water or mountain air.

Sex is not everything, of course, but it is a catalyst for many other things and, since so many other things must be right for it to function well, also a touchstone for the quality of the total relationship. When it is good people look different. The emotional atmosphere one senses in a house where it is right is one of calm and peace, yet also of lightness, fun and humor, and everything moves easily. Above all the children sense it and are happy for it, though they do not necessarily know what they sense, any more than they know, or care, about the cost of the good food that nourishes them.

A. C. ROBIN SKYNNER, *One Flesh; Separate Persons*, 1976

From pent-up, aching rivers;
From that of myself, without which I were nothing;
From what I am determined to make illustrious, even if I stand
  sole among men;
From my own voice resonant—singing the phallus,
Singing the song of procreation,
Singing the need of superb children, and therein superb grown
  people,
Singing the muscular urge and the blending,
Singing the bedfellow's song, (O resistless yearning!
O for any and each, the body correlative attracting!
O for you, whoever you are, your correlative body! O it, more
  than all else, you delighting!)
—From the hungry gnaw that eats me night and day;
From native moments—from bashful pains—singing them;
Singing something yet unfound, though I have diligently sought
  it, many a long year,
Singing the true song of the Soul, fitful, at random;
Singing what, to the Soul, entirely redeemed her, the faithful one,
  even the prostitute, who detained me when I went to the city;
Singing the song of prostitutes;
Renascent with grossest Nature, or among animals;
Of that—of them, and what goes with them, my poems
  informing;
Of the smell of apples and lemons—of the pairing of birds,

Of the wet of woods—of the lapping of waves,
Of the mad pushes of waves upon the land—I them chanting;
The overture lightly sounding—the strain anticipating;
The welcome nearness—the sight of the perfect body;
The swimmer swimming naked in the bath, or motionless on his back lying and floating;
The female form approaching—I, pensive, love-flesh tremulous, aching;
The divine list, for myself or you, or for any one, making;
The face—the limbs—the index from head to foot, and what it arouses;
The mystic deliria—the madness amorous—the utter abandonment;
(Hark close, and still, what I now whisper to you,
I love you—O you entirely possess me,
O I wish that you and I escape from the rest, and go utterly off—O free and lawless,
Two hawks in the air—two fishes swimming in the sea not more lawless than we;)
—The furious storm through me careering—I passionately trembling;
The oath of the inseparableness of two together—of the woman that loves me, and whom I love more than my life—that oath swearing;
(O I willingly stake all, for you!
O let me be lost, if it must be so!
O you and I—what is it to us what the rest do or think?
What is all else to us? only that we enjoy each other, and exhaust each other, if it must be so:)
—From the master—the pilot I yield the vessel to;
The general commanding me, commanding all—from him permission taking;
From time the programme hastening, (I have loitered too long, as it is;)
From sex—From the warp and from the woof;
(To talk to the perfect girl who understands me,
To waft to her these from my own lips—to effuse them from my own body;)
From privacy—from frequent repinings alone;
From plenty of persons near, and yet the right person not near;

From the soft sliding of hands over me, and thrusting of fingers
through my hair and beard;
From the long sustained kiss upon the mouth or bosom;
From the close pressure that makes me or any man drunk,
fainting with excess;
From what the divine husband knows—from the work of
fatherhood;
From exultation, victory, and relief—from the bedfellow's
embrace in the night;
From the act-poems of eyes, hands, hips, and bosoms,
From the cling of the trembling arm,
From the bending curve and the clinch,
From side by side, the pliant coverlid off-throwing,
From the one so unwilling to have me leave—and me just as
unwilling to leave,
(Yet a moment, O tender waiter, and I return;)
—From the hour of shining stars and dropping dews,
From the night, a moment, I, emerging, flitting out,
Celebrate you, act divine—and you, children prepared for,
And you, stalwart loins.

WALT WHITMAN, 'Children of Adam', 1881

Behold, thou art fair, my love; behold, thou art fair; thou hast doves' eyes within thy locks: thy hair is as a flock of goats, that appear from mount Gilead.

Thy teeth are like a flock of sheep that are even shorn, which came up from the washing; whereof every one bear twins, and none is barren among them.

Thy lips are like a thread of scarlet, and thy speech is comely: thy temples are like a piece of a pomegranate within thy locks.

Thy neck is like the tower of David builded for an armoury, whereon there hang a thousand bucklers, all shields of mighty men.

Thy two breasts are like two young roes that are twins, which feed among the lilies.

Until the day break, and the shadows flee away, I will get me to the mountain of myrrh, and to the hill of frankincense.

Thou art all fair, my love; there is no spot in thee.

Come with me from Lebanon, my spouse, with me from Lebanon:

look from the top of Amana, from the top of Shenir and Hermon, from the lions' dens, from the mountains of the leopards.

Thou hast ravished my heart, my sister, my spouse; thou hast ravished my heart with one of thine eyes, with one chain of thy neck.

How fair is thy love, my sister, my spouse! how much better is thy love than wine! and the smell of thine ointments than all spices!

Thy lips, O my spouse, drop as the honeycomb: honey and milk are under thy tongue; and the smell of thy garments is like the smell of Lebanon.

A garden inclosed is my sister, my spouse; a spring shut up, a fountain sealed.

Thy plants are an orchard of pomegranates, with pleasant fruits; camphire, with spikenard,

Spikenard and saffron; calamus and cinnamon, with all trees of frankincense; myrrh and aloes, with all the chief spices:

A fountain of gardens, a well of living waters, and streams from Lebanon.

Awake, O north wind; and come, thou south; blow upon my garden, that the spices thereof may flow out. Let my beloved come into his garden, and eat his pleasant fruits.

The Song of Solomon

6

Like Olive Plants around the Table

'THY wife shall be as a fruitful vine by the sides of thine house; thy children like olive plants round about thy table' (Psalm 128).

The psalmist paints a deceptively peaceful picture of children in the home. Even in the days when children were 'seen but not heard' they can scarcely have been so quiet or passive.

After the arrival of the first child a marriage changes crucially. The twosome turns into a threesome, producing new strains and tensions. Husband and wife have to learn to be father and mother, and they may not both take to their new roles with equal ease:

> His feelings for this little creature were not at all what he had expected. There was not an atom of pride or joy in them; on the contrary, he was oppressed by a new sense of apprehension—the consciousness of another vulnerable region.
>
> (TOLSTOY, *Anna Karenina*)

The intensity and exaltation, as well as the anguish, of a shared experience of birth contrasts with the desolation of a miscarriage or a still-birth. And childlessness is itself a factor in a marriage. The death of a child brings some couples, like the Darwins, closer together, but it can also cause a rift: grief and desolation are hard to share.

Children may, as Strindberg says, be the cement of a marriage, but they can also be used as hostages and weapons, or get caught up in the parents' strife, and become, like Henry James's Maisie, 'a little feathered shuttlecock they could fiercely keep flying between them'.

'The Joyes of Parents are Secret; and so are their Griefes and Feares', wrote Francis Bacon, but the joys (like the joys of sex in marriage) are not chronicled nearly as often as the griefs. As children grow older, parents may have to accept that the centre of the stage is no longer theirs: 'Something is pushing them / To the side of their own lives' (Philip Larkin, 'Afternoons'). And when children reach adolescence, all hell can break loose, for the parents as well as the children. Grown-ups may be dragged back into the turmoils of adolescence, a state of mind they thought they had long since outgrown. Like Alison Lurie's couple, they may flail about like confused and stroppy youngsters themselves.

Allowing one's children to separate and go their own way can also be hard. Parents can feel painfully bereft when children leave home, and nostalgia for one's children's childhood is probably more common than for one's own:

> she would feel her throat swell at the memory of banana sandwiches and strawberries, and in the winter the smell of warm buttered toast almost made her cry. She had never spoken to anyone of this babyish trait, though sometimes she wondered whether it was her own childhood that she mourned, or the childhood of her children.
>
> (ALICE THOMAS ELLIS, *The Other Side of the Fire*).

Yet the notion that a child is a guest in the house, to be loved and respected, but never possessed, is a truth as old as the sacred scripts of the Vedanta.

Whoever considers the length and feebleness of human infancy, with the concern which both sexes naturally have for their offspring, will easily perceive, that there must be an union of male and female for the education of the young, and that this union must be of considerable duration.

DAVID HUME, *A Treatise of Human Nature*, 1740

Since marriage, which is a human institution invented for purely practical purposes, is so frail and so full of stumbling-blocks, how is it that so many marriages hold together? They do so because both partners have one interest in common, the thing for which nature has always intended marriage, namely children. Man is in a state of perpetual conflict with nature, in which he is perpetually being vanquished. Take two lovers who want to live together, partly in order to enjoy themselves, partly for the sake of being in each other's company. They regard any talk of possible children as an insult. Long before a child arrives they discover that their bliss is not so heavenly after all, and their relationship becomes stale. Then a child is born. Everything is new again and now, for the first time, their relationship is beautiful, for the ugly egoism of the duet has vanished.... Children are what holds a marriage together.

AUGUST STRINDBERG, Preface to *Getting Married*, 1884

Parenthood does not begin with the birth of the child, nor with the germination of the seed. It has its beginnings even before the child's conception, and is part and parcel of the marriage and the love from the expression of which the baby owes his being. If this is so our abilities as good mothers and fathers—and our failures too—cannot be viewed in isolation, apart from the sum total—the interaction and fusing of personalities—which makes up the marriage.

SHEILA KITZINGER, *The Experience of Childbirth*, 1962

They told me that the Erewhonians believe in pre-existence; and not only this . . . but they believe that it is of their own free act and deed in a previous state that they come to be born into this world at all. They hold that the unborn are perpetually plaguing and tormenting the married of both sexes, fluttering about them incessantly, and giving them no peace either of mind or body until they have consented to take them under their protection. If this were not so (this at least is what they urge), it would be a monstrous freedom for one man to take with another, to say that he should undergo the chances and changes of this mortal life without any option in the matter. No man would have any right to get married at all, inasmuch as he can never tell what frightful misery his doing so may entail forcibly upon a being who cannot be unhappy as long as he does not exist.

SAMUEL BUTLER, *Erewhon*, 1872

Ethel and Bernard returned from their Honymoon with a son and hair a nice fat baby called Ignatius Bernard. They soon had six more children four boys and three girls and some of them were twins which was very exciting.

DAISY ASHFORD, *The Young Visiters*, 1919

15 June 1858

What you say of the pride of giving life to an immortal soul is very fine, dear, but I own I cannot enter into that; I think much more of our being like a cow or a dog at such moments; when our poor nature becomes so very animal and unecstatic—but for you, dear, if you are sensible and reasonable not in ecstasy nor spending your day with nurses and wet nurses, which is the ruin of many a refined and intellectual young lady, without adding to her real maternal duties, a child will be a great resource.

QUEEN VICTORIA, letter to Princess Frederick William

BIRTH

Did my hand ever touch your hair?
Did my fingers ever feel your soft skin?

Always between us a winter frost descended,
A summer haze drifted, didn't it?

Yet your belly, swollen with child,
Twitches and jumps with the quickening.

We slept between the same sheets, yet
I don't know who you are.
The child you will bear could be you
Or it might well be me.

And you, as well, don't know who I am.

Now you hold two lives: I can no more see you as an individual.
All your love is centred now on the child you will bear.
I, the father—my being is less important than the child's rearing.

This is illusion, perhaps:
Neither you nor I really exist:
Merely, through the sense of touch, at times we feel alive
Just as in a dream,
And our child will be born to this dream.

What's born doesn't live for ever
And what exists isn't reborn.
So, whether you exist or are something that's been born,
I do not know.

TAKASHI SHINKICHI (1901–)

Of all the riddles of a married life, said my father, crossing the landing in order to set his back against the wall, whilst he propounded it to my uncle Toby—of all the puzzling riddles, said he, in a marriage state,—of which you may trust me, brother Toby, there are more asses loads than all Job's stock of asses could have carried—there is not one that has more intricacies in it than this—that from the very moment the mistress of the house is brought to bed, every female in it, from my lady's gentlewoman down to the cinder-wench, becomes an inch taller for it; and give themselves more airs upon that single inch, than all their other inches put together.

I think rather, replied my uncle Toby, that 'tis we who sink an inch lower.—If I meet but a woman with child—I do it.—'Tis a heavy tax upon that half of our fellow-creatures, brother Shandy, said my uncle Toby—'Tis a piteous burden upon 'em, continued he, shaking his head—Yes, yes, 'tis a painful thing—said my father, shaking his head too—but certainly since shaking of heads came into fashion, never did two heads shake together, in concert, from two such different springs.

God bless } 'em all——said my uncle Toby and my father,
Deuce take } each to himself.

LAURENCE STERNE, *The Life and Opinions of Tristram Shandy*, 1759–67

'And did you really,' says she, 'make your wife's caudle yourself?'

'Indeed, madam,' said he, 'I did; and do you think that so extraordinary?'

'Indeed I do,' answered she; 'I thought the best husbands had looked on their wives' lying-in as a time of festival and jollity. What! did you not even get drunk in the time of your wife's delivery? tell me honestly how you employed yourself at this time.'

'Why, then, honestly,' replied he, 'and in defiance of your laughter, I lay behind her bolster, and supported her in my arms; and upon my soul, I believe I felt more pain in my mind than she underwent in her body. And now answer me as honestly: Do you really think it a proper time of mirth, when the creature one loves to distraction is undergoing the most racking torments, as well as in the most imminent danger? and—but I need not express any more tender circumstances.'

HENRY FIELDING, *Amelia*, 1751

Now Julie kicked the bedclothes clear of her feet. She lifted her nightdress and turned and crouched on all fours. She spread her elbows until her face was pressed into the pillows. He murmured her name at the sight—in a body so dignified and potent—of the sweet helplessness of her raised buttocks, untidily framed by the embroidered hem of her nightie. The silence resounded after all their promises, and merged with the stirring of a billion needles in the plantation. He moved inside her gently. Something was gathering up around them, growing louder, tasting sweeter, getting warmer, brighter, all senses were synthesising, condensing in the idea of increase. She called out quietly, over and again, drawn out sounds of 'oh', each of which dipped and rose in pitch like a baffled question. Later, she shouted something joyful he could not make out, lost as he was to meaning. Then she was pulling away from him, she wanted to be on her back. She settled herself squarely and drew a sharp breath. She rested the tips of the fingers of one hand on the lower part of her belly and massaged herself lightly. He remembered the pretty name, effleurage. With the other hand she clasped him, squeezing tighter as the contraction gained in intensity, in this way communicating its progress. She was prepared, she was controlling her breathing, making steady, rhythmic exhalations that accelerated into shallow panting as she approached the peak. She was off on this second journey alone, all he could do was run along the shore and call encouragement. She was going from him, lost to the process. Her fingers dug into his hand. His pulse was banging in his temples, disturbing his vision. He tried to keep the fear out of his voice. He had to remember his lines. 'Ride it, ride this wave, don't fight it, float with it, float . . . ' Then he joined in her panting, making a heavy emphasis on the breathing out, slowing down as the grip on his hand loosened. He suspected that the form of his participation had been devised by medical authorities to oppose the panic of paternal helplessness.

As the contraction passed they took a deep breath together. Julie cupped her hands over her mouth to counter the sickness of hyperventilation. She said something, but her words were muffled. He waited. She dropped her hands and smiled wryly. They returned to the room, and to themselves, as though emerging from shelter after a storm. He could not recall what they had been talking about, or whether they had been talking. It did not matter.

'Do you remember it all?' Julie said. She was not asking him to reminisce. She wanted to know if he knew what to do.

He nodded. He would have liked to take a peek at one of Julie's books. There were precise stages of labour, as he half remembered, different breathing techniques associated with them, time to hold back, time when it was important to let go. But there was a long day ahead. There would be leisure for that. And he remembered the last time clearly enough. He had been brow mopper, telephonist, flower man, champagne pourer, midwife's dogsbody, and he had talked her through. Afterwards she had told him he had been useful. His impression was his value had been more symbolic. He dressed, then crossed the room and found a pair of Julie's socks to put on.

'Where's the midwife's number?'

'In my coat pocket, hanging behind the door. Put the kettle on as you go out. Make two hot-water bottles when you come back. And a pot of jasmine tea. Both the fires need building up.' He remembered too these husky commands, the mother's absolute right to order her own domain.

Outside the dawn was still confined to the eastern sky. The clouds had disappeared entirely and for the first time he saw stars. The moon was still the main source of light. He walked quickly up the brick path in his wet shoes, noticing that Julie had taken the precaution of sweeping the snow clear. The phone box on the corner had no light inside and he had to feel for the numbers. When he got through, he found he was talking to a receptionist in a medical centre in the nearby town. He was not to worry. The midwife would be contacted, and would arrive within the hour.

On the way back, as he walked the short stretch of road he had run along less than an hour before, he slowed and tried to take the measure of the transformations; but he was incapable of reflection; he could think only of details, of tea, logs, and hot-water bottles.

The cottage was quiet when he returned. He prepared the tea tray, fetched wood from a lean-to outside, built up the downstairs fire and filled a basket for the one upstairs. He scanned Julie's shelves without success for books on birth. To buck himself up with a show of competence, he stood at the kitchen sink for several minutes scrubbing his hands.

Balancing the tray on the basket, and holding the hot-water bottles under his arm, he tottered upstairs. Julie was sprawled on her back. Her hair was damp and clung to her neck and forehead. She was agitated, querulous.

'You said you wouldn't be long. What have you been doing?'

He was about to dispute with her when he remembered that irritability could be part of the process, one of the markers along the route. But surely that should have come later. Had they missed out some stages? He gave her the tea and offered her a massage. She could not bear to be touched, however. He arranged the bedclothes for her. Recalling how furious she had been before when the midwife had spoken to her like a child, he adopted the tone of a soft-spoken football coach.

'Move your leg, this way. Good. Everything's looking good. We're on course.' And so on. She was not really mollified, but she complied, and she drank the tea.

He was blowing on the embers, encouraging a flame across a handful of twigs, when he heard her call his name. He hurried over. She was shaking her head. She made as if to place her fingers over her belly, and then gave up.

'I've been up all night. I'm too tired for this, I'm not ready.'

His words of encouragement were cut off by a long shout. She fought to inhale, and there was another, a prolonged hoot of astonishment.

'Ride it out, ride the wave . . . ' he began to say. Again, his words were cut off. He had lost his place. Exhortations to rhythmic breathing were now inane. A gale had torn the instructions away from him. She held his forearm with both hands in a fierce grip. Her teeth were bared, the muscles and tendons of her neck were stretching to breaking point. He was lost. He could give her nothing more than his forearm.

He called out to her, 'Julie, Julie, I'm here with you.'

But she was alone. She was drawing breath and shouting again, this time wildly, as though in exhilaration, and when she had no more air in her lungs, it did not matter, the shout had to go on, and on. The contraction lifted her off her back and twisted her on to her side. The sheet was still gathered up to her waist and had knotted itself round her. He felt the bed frame tremble with her effort. There was a final click at the back of her throat and she was drawing breath again, tossing her head as she did so. When she looked at him, past him, her eyes were bright, wide with purpose. The brief despair had gone. She was back in control. He thought she was about to speak, but the grip on his arm was tightening again and she was away. Her lips shivered as they stretched tighter over her teeth and from deep in her chest came a strangled groan, a bottled-up, gurgling sound of colossal, straining effort. Then it tapered away, and she let her head fall back against the pillows.

She took deep breaths and spoke in a surprisingly normal voice. 'I need a cold drink, a glass of water.' He was about to stand when she restrained him. 'But I don't want you to go away. I think it might be coming.'

'No, no. The midwife isn't here yet.'

She smiled as if he had made a joke for her benefit. 'Tell me what you can see.'

He had to reach under her to get the sheet clear.

There was a shock, a jarring, a slowing down as he entered dream time. A quietness enveloped him. He had come before a presence, a revelation. He was staring down at the back of a protruding head. No other part of the body was visible. It faced down into the wet sheet. In its silence and complete stillness there was an accusation. Had you forgotten me? Did you not realise it was me all along? I am here. I am not alive. He was looking at the whorl of wet hair about the crown. There was no movement, no pulse, no breathing. It was not alive, it was a head on the block, and yet the demand was clear and pressing. This was my move. Now what is yours? Perhaps a second had passed since he had lifted back the sheet. He put out his hand. It was a blue-white marble sculpture he was touching, both inert and full of intent. It was cold, the wetness was cold, and beneath that there was a warmth, but too faint, the residual, borrowed warmth of Julie's body. That it was suddenly and obviously there, a person not from another town or from a different country, but from life itself, the simplicity of that, was communicating to him a clarity and precision of purpose. He heard himself say something reassuring to Julie, while he himself was comforted by a memory, brief and clear like a firework, of a sunlit country road, of wreckage and a head. His thoughts were resolving into simple, elementary shapes. This is really all we have got, this increase, this matter of life loving itself, everything we have has to come from this.

Julie was not yet ready to push. She was recovering her strength. He slid his hand round to the face, found the mouth and used his little finger to clear it of mucus. There was no breath. He moved his fingers down, below the lip of Julie's taut skin, to find the hidden shoulder. He could feel the cord there, thick and robust, a pulsing creature wound twice in a noose about the neck. He worked his forefinger round and pulled cautiously. The cord came easily, copiously, and as he lifted it clear of the head, Julie gave birth—he saw in an instant how active and generous the verb was—she summoned her will and her physical strength and gave. With a creaking, waxy sound the child slid into his hands. He saw only the long back, powerful and slippery, with grooved, muscular spine. The cord, still beating, hung across the shoulder and tangled round a foot. He was only the catcher, not the home, and his one thought was to return the child to its mother. As he was lifting it across they heard a snuffling sound and a single lucid cry. It lay face down with an ear towards its mother's heart. They drew the covers over it. Because the hot-water bottles were too heavy and hot, Stephen climbed into bed beside Julie and they kept the baby warm between them. The breathing was settling into a rhythm, and a warmer colour, a bloom of deep pink, was suffusing its skin.

It was only then that they began to exclaim and celebrate, and kiss and nuzzle the waxy head which smelled like a freshly baked bun. For minutes they were beyond forming sentences and could only make noises of triumph and wonder, and say each other's names aloud. Anchored by its cord, the baby lay with its head resting between its closed fists. It was a beautiful child. Its eyes were open, looking towards the mountain of Julie's breast. Beyond the bed was the window through which they could see the moon sinking into a gap in the pines. Directly above the moon was a planet. It was Mars, Julie said. It was a reminder of a harsh world. For now, however, they were immune, it was before the beginning of time, and they lay watching planet and moon descend through a sky that was turning blue.

They did not know how much later it was they heard the midwife's car stop outside the cottage. They heard the slam of its door and the tick of hard shoes on the brick path.

'Well?' Julie said. 'A girl or a boy?' And it was in acknowledgement of the world they were about to rejoin, and into which they hoped to take their love, that she reached down under the covers and felt.

IAN MCEWAN, *The Child in Time*, 1987

INFANT SORROW

My mother groaned, my father wept:
Into the dangerous world I leapt,
Helpless, naked, piping loud,
Like a fiend hid in a cloud.

Struggling in my father's hands,
Striving against my swaddling bands,
Bound and weary, I thought best
To sulk upon my mother's breast.

WILLIAM BLAKE, *Songs of Experience*, 1794

She was born in the autumn and was a late fall in my life, and lay purple and dented like a little bruised plum, as though she'd been lightly trodden in the grass and forgotten.

Then the nurse lifted her up and she came suddenly alive, her bent legs kicking crabwise, and her first living gesture was a thin wringing of the hands accompanied by a far-out Hebridean lament.

This moment of meeting seemed to be a birthtime for both of us; her first and my second life. Nothing, I knew, would be the same again, and I think I was reasonably shaken. I peered intently at her, looking for familiar signs, but she was convulsed as an Aztec idol. Was this really my daughter, this purple concentration of anguish, this blind and protesting dwarf?

Then they handed her to me, stiff and howling, and I held her for the first time and kissed her, and she went still and quiet as though by instinctive guile, and I was instantly enslaved by her flattery of my powers.

LAURIE LEE, *I Can't Stay Long*, 1975

Mon très cher Père! Vienne, *ce* 18 *de Juin*, 1783

Congratulations, you are a grandpapa! Yesterday, the 17th, at half past six in the morning my dear wife was safely delivered of a fine sturdy boy, as round as a ball. Her pains began at half past one in the morning, so that night we both lost our rest and sleep. At four o'clock I sent for my mother-in-law—and then for the midwife. At six o'clock the child began to appear and at half past six the trouble was all over. . . .

My dear wife, who kisses your hands and embraces my dear sister most affectionately, is as well as she can be in the circumstances. I trust with God's help that, as she is taking good care of herself, she will make a complete recovery from her confinement.

Mon très cher Père! Vienne, *ce 21 de Juin*, 1783

This will have to be a very short letter. I must only tell you what is absolutely necessary, as I have far too much to do. For a new Italian opera is being produced, in which for the first time two German singers are appearing, Madame Lange, my sister-in-law, and Adamberger, and I have to compose two arias for her and a rondo for him. I hope you received my last letter of rejoicing. Thank God, my wife has now survived the two critical days, yesterday and the day before, and in the circumstances is very well. We now hope that all will go well. The child too is quite strong and healthy and has a tremendous number of things to do, I mean, drinking, sleeping, yelling, pissing, shitting, dribbling and so forth. He kisses the hands of his grandpapa and of his aunt.

Mon très cher Père, Vienne, *ce 5 de Juillet*, 1783

We both thank you for the prayer you made to God for the safe delivery of my wife. Little Raimund is so like me that everyone immediately remarks it. It is just as if my face had been copied. My dear little wife is absolutely delighted, as this is what she had always desired. He will be three weeks old next Tuesday and he has grown in an astonishing manner. As for the opera . . .

WOLFGANG AMADEUS MOZART, letters to his father

Claudia thought that she must look hugely complacent and at home in this position; that they must present all the appearance of a couple, and then remembered that when she'd had her first baby she had realised with astonishment that the perfect couple consisted of a mother and child and not, as she had always supposed, a man and woman.

ALICE THOMAS ELLIS, *The Other Side of the Fire*, 1983

Several times in those years Amy Parker attempted to have their child, but evidently this was not intended to happen.

'This is a barren stretch of the road,' she said, laughing.

For nothing was coming out of Quigleys or O'Dowds, and now Parkers were adopting the evasions and pretences of a childless intimacy. They had persuaded themselves that their neat house, which Stan and the Quigley boys had built, was not the box which enclosed their lives. They were still young, of course, so that their fallibility had not yet been revealed, except by flashes, which can be dismissed as dreams. Even though circumstances had started them to think, it was in a tangled way, in which they made little progress against the knots of thought. They were praying too, more or less regularly, in accordance with the fluctuations of belief. They loved, sometimes with inspiration, also occasionally with resentment. They desired each other's presence perhaps less than before, cherishing the moments of peace, even of past sorrow. Sometimes they made excuses for each other.

'We can get along all right as we are,' said Stan Parker. 'If you have kids, they can blame you for it forever after.'

It was like that.

PATRICK WHITE, *The Tree of Man*, 1955

NOT TO BE BORN

No different, I said, from rat's or chicken's,
That ten-week protoplasmic blob. But you
Cried as if you knew all that was nonsense
And knew that I did, too.

Well, I had to say something. And there
Seemed so little anyone could say.
That life had been in women's wombs before
And gone away?

This was our life. And yet, when the dead
Are mourned a little, then become unreal,
How should the never born be long remembered?
So this in time will heal

Though now I cannot comfort. As I go
The doctor reassures: 'Straightforward case.
You'll find, of course, it leaves her rather low.'
Something is gone from your face.

DAVID SUTTON, 1982

A still-birth, and even more the death of a child, can tear a marriage apart. 'It does not even seem to me to have drawn me nearer to Maud', wrote A. C. Benson after the death of his son. 'My power of loving seems extinguished.' Other couples can share their grief and remain close to one another:

My feelings of longing after our lost treasure make me feel painfully indifferent to the other children but I shall get right in my feelings to them before long. You must remember that you are my prime treasure (and always have been) my only hope of consolation is to have you safe home and weep together. I feel so full of fears about you, they are not reasonable fears but my power of hoping seems gone. I hope you will let dearest Fanny or Catherine, if she comes, stay with you til the end. I can't bear to think of you by yourself. . . . Your letter has just come. Do not be in a hurry to set off. You do give me the only comfort I can take in thinking of her happy innocent life. She never concealed a thought and so affectionate and forgiving. What a blank it is. Don't think of coming in one day. We shall be much less miserable together.

¶ *From a letter from Emma Darwin, pregnant with their ninth child, to her husband Charles, who was taking care of their dying 10-year-old daughter Annie.*

Charles's letter announcing the little girl's death followed shortly after:

My dearest Emma,

I pray God Fanny's note may have prepared you. She went to her final sleep most tranquilly, most sweetly at 12 o'clock today. Our poor dear child had had a very short life but I trust happy and God only knows what miseries might have been in store for her. She expired without a sigh. How desolate it makes one to think of her frank cordial manners. I am so thankful for the daguerrotype. I cannot remember ever seeing the dear child naughty, God bless her. We must be more and more to each other my dear wife—Do what you can to bear up and think how invariably kind and tender you have been to her—I am in bed not very well with my stomach. When I shall return I cannot yet say. My own poor dear dear wife.

He that sincerely loves his Wife and Family, and studies to improve that Affection in himself, conceives Pleasure from the most indifferent things; while the married Man who has not bid adieu to the Fashions and false Gallantries of the Town, is perplexed with every thing around him. In both these Cases Men cannot, indeed, make a sillier Figure, than in repeating such Pleasures and Pains to the rest of the World; but I speak of them only, as they sit upon those who are involved in them. As I visit all Sorts of People, I cannot indeed but smile, when the good Lady tells her Husband what extraordinary things the Child spoke since he went out. No longer than Yesterday I was prevailed with to go home with a fond Husband; and his Wife told him, that his Son, of his own Head, when the Clock in the Parlour struck Two, said Pappa would come home to Dinner presently. While the Father has him in a Rapture in his Arms, and is drowning him with Kisses, the Wife tells me he is but just four Year old. Then they both struggle for him, and bring him up to me, and repeat his Observation of two a Clock. I was called upon by Looks upon the Child, and then at me, to say something; and I told the Father, that this Remark of the Infant of his coming home, and joyning the Time with it, was a certain Indication that he would be a great Historian and Chronologer. They are neither of them Fools, yet received my Compliment with great Acknowledgment of my Prescience. I fared very well at Dinner, and heard many other notable Sayings of their Heir, which would have given very little Entertainment to one less turn'd to Reflection than I was; but it was a pleasing Speculation to remark on the Happiness of a Life, in which Things of no Moment give Occasions of Hope, Self-Satisfaction, and Triumph. On the other Hand, I have known an ill-natur'd Coxcomb, who was hardly improved in any thing but Bulk, for want of this Disposition, silence the whole Family, as a Set of silly Women and Children, for recounting Things which were really above his own Capacity.

The Spectator, no. 479, 9 September 1712

When I consider how little of a rarity children are—that every street and blind alley swarms with them,—that the poorest people commonly have them in most abundance,—that there are few marriages that are not blest with at least one of these bargains,—how often they turn out ill, and defeat the fond hopes of their parents taking to vicious courses, which end in poverty, disgrace, the gallows etc.—I cannot for my life tell what cause for pride there can possibly be in having them. If they were young phœnixes, indeed, that were born but one in a year, there might be a pretext. But when they are so common—

I do not advert to the insolent merit which they assume with their husbands on these occasions. Let them look to that. But why *we*, who are not their natural-born subjects, should be expected to bring our spices, myrrh, and incense,—our tribute and homage of admiration, I do not see.

'Like as the arrows in the hand of the giant, even so are the young children:' so says the excellent office in our Prayer-book appointed for the churching of women. 'Happy is the man that hath his quiver full of them:' So say I; but then don't let him discharge his quiver upon us that are weaponless;—let them be arrows, but not to gall and stick us. I have generally observed that these arrows are double-headed; they have two forks, to be sure to hit with one or the other. As for instance, where you come into a house which is full of children, if you happen to take no notice of them (you are thinking of something else, perhaps, and turn a deaf ear to their innocent caresses), you are set down as untractable, morose, a hater of children. On the other hand, if you find them more than usually engaging,—if you are taken with their pretty manners, and set about in earnest to romp and play with them, some pretext or other is sure to be found for sending them out of the room: they are too noisy or boisterous, or Mr — does not like children. With one or other of these forks the arrow is sure to hit you.

I could forgive their jealousy, and dispense with toying with their brats, if it gives them any pain; but I think it unreasonable to be called upon to *love* them, where I see no occasion,—to love a whole family, perhaps eight, nine, or ten, indiscriminately,—to love all the pretty dears, because children are so engaging.

I know there is a proverb, 'Love me, love my dog:' that is not always so very practicable, particularly if the dog be set upon you to tease you or snap at you in sport. But a dog, or a lesser thing,—any inanimate substance, as a keepsake, a watch or a ring, a tree, or the place where we last parted when my friend went away upon a long absence, I can make shift to love, because I love him, and anything that reminds me of him; provided it be in its nature indifferent, and apt to receive whatever hue fancy can give it. But children have a real character and an essential being of themselves: they are amiable or unamiable *per se*; I must love or hate them as I see cause for either in their qualities. A child's nature is too serious a thing to admit of its being regarded as a mere appendage to another being, and to be loved or hated accordingly: they stand with me upon their own stock, as much as men and woman do. O! but you will say, sure it is an attractive age,—there is something in the tender years of infancy that of itself charms us. That is the very reason why I am more nice about them. I know that a sweet child is the sweetest thing in nature, not even excepting the delicate creatures which bear them; but the prettier the kind of a thing is, the more desirable it is that it should be pretty of its kind. One daisy differs not much from another in glory; but a violet should look and smell the daintiest.—I was always rather squeamish in my women and children.

CHARLES LAMB, *Essays of Elia*, 1823

DID SOMEONE SAY 'BABIES'?

Everybody who has a baby thinks everybody who hasn't a baby
ought to have a baby,
Which accounts for the success of such plays as the Irish Rose of
Abie,
The idea apparently being that just by being fruitful
You are doing something beautiful,
Which if it is true
Means that the common housefly is several million times more
beautiful than me or you.
Also, everybody who hasn't a baby thinks it correct to give
tongue
To ecstatic phrases and clauses at the sight of other people's
young.
Who is responsible for this propaganda that fills all our houses
from their attics to their kitchens?
Is it the perambulator trust or the safety pin manufacturers or the
census takers or the obstetritchens?
Why do we continue not only to be hoodwinked by them but
even lend ourselves to furthering their plots
By all the time talking about how nice it is to have a houseful of
tots?

Men and women everywhere would have a lot more chance of
acquiring recreation and fame and financial independence
If they didn't have to spend most of their time and money
tending and supporting two or three unattractive
descendants.
We could soon upset this kettle of fish, forsooth,
If every adult would only come out and tell every other adult the
truth.
To arms, adults! Kindle the beacon fires!
Women, do you want to be nothing but dams? Men, do you
want to be nothing but sires?
To arms, Mr President! Call out the army, the navy, the marines,
the militia, the cadets and the middies.
Down with the kiddies!

OGDEN NASH, 1941

After Sam was born, I remember thinking that no one had ever told me how much I would love my child; now, of course, I realized something else no one tells you: that a child is a grenade. When you have a baby, you set off an explosion in your marriage, and when the dust settles, your marriage is different from what it was. Not better, necessarily; not worse, necessarily; but different. All those idiotically lyrical articles about sharing child-rearing duties never mention that, nor do they allude to something else that happens when a baby is born, which is that all the power struggles of the marriage have a new playing field. The baby wakes up in the middle of the night, and instead of jumping out of bed, you lie there thinking: Whose turn is it? If it's your turn, you have to get up; if it's his turn, then why is he still lying there asleep while you're awake wondering whose turn it is? Now it takes *two* parents to feed the child—one to do it and one to keep the one who does it company. Now it takes two parents to take the child to the doctor—one to do it and one to keep the one who does it from becoming resentful about having to do it. Now it takes two parents to fight over who gets to be the first person to introduce solids or the last person to notice the diaper has to be changed or the one who cares most about limiting sugar snacks or the one who cares least about conventional discipline.

No one ever tells you these things—not that we would have

listened had anyone tried. We were so smart. We were so old. We were so happy. We had it knocked.

NORA EPHRON, *Heartburn*, 1983

Brian and Erica, like their friends, students, and colleagues, have spent considerable time trying to understand and halt the war in Vietnam. If he were to draw a parallel between it and the war now going on in his house, he would have unhesitatingly identified with the South Vietnamese. He would have said that the conflict, begun a year or so ago as a minor police action, intended only to preserve democratic government and maintain the *status quo*—a preventive measure, really—has escalated steadily and disastrously against his and Erica's wishes, and in spite of their earnest efforts to end it. For nearly two years, he would point out, the house on Jones Creek Road has been occupied territory. Jeffrey and Matilda have gradually taken it over, moving in troops and supplies, depleting natural resources, and destroying the local culture.

From the younger Tates' position, however, the parallel is reversed. Brian and Erica are the invaders: the large, brutal, callous Americans. They are vastly superior in material resources and military experience, which makes the war deeply unfair; and they have powerful allies like the Corinth Public School System. The current position of Jeffrey and Matilda is, from their own viewpoint, almost tragic. In spite of their innate superiority and their wish for self-government, they remain dependent on Erican aid and Brian Tate's investments. Worse still in some ways is the barrage of propaganda and lies they have to endure. Brian and Erica keep insisting publicly that they are not trying to destroy Jeffrey or Matilda, but instead fighting to preserve the best, the most enlightened and democratic elements within them. When they hear these lies the younger Tates naturally feel exploited and furious. They refuse to negotiate and retreat into the jungles of their rooms on the third floor, where they plan guerrilla attacks.

Brian and Erica are at a moral and psychological disadvantage in the war because they want to save face both at home and abroad. They have been favoured by environment and heredity, and wish to think themselves worthy of their good fortune. They therefore desire (against all reason) to enjoy the affection, respect, and gratitude of the people they are at war with and whose territory they have invaded;

and they never cease to be deeply hurt and indignant that they are not receiving this affection, etc.

Externally, too, Brian and Erica have a reputation to uphold. For many years they have been generally regarded, and they have regarded themselves, as democratic peace- and freedom-loving persons, devoted to decent and humanitarian goals. So great was their need to preserve this reputation that they never declared war officially, but continued to speak of the conflict as a peace-keeping effort and to insist that they were acting in an advisory capacity. Nevertheless, the true facts are widely known, and have earned them the bad opinion of the rest of the world—including that of other parents who are currently engaged in their own undeclared wars.

Jeffrey and Matilda, on the other hand, do not have to worry about public opinion. They know they are right. They know that any belligerent action they might undertake will be applauded by their contemporaries, some of whom have already gone much further in terms of overt hostility. The magazines they read, the songs they hear, their whole culture supports them. Even on the enemy side there are many who dare to take their part, repudiating natural adult allegiances in the cause of revolution and truth.

As yet, Brian has won most of the pitched battles; but the effort of winning is exhausting his resources, and he knows it. He knows that time is against him, and he cannot win the war. Even now his victories are all negative ones: he has, once more, beaten off an attack on some stronghold, or contained the enemy within the existing combat zone. He can, for example, wearily congratulate himself on the fact that his children do not—as far as he knows—steal cars or bomb buildings or inject themselves with drugs; that they have not got themselves arrested by the police yet, or pregnant. Sometimes Brian wishes they had done so; then at least they would be somewhere else—in jail or an unwed mother's home—and someone else would be responsible for them.

What makes the war most exhausting for Brian is that his ally, Erica, has deserted him. She has declared, not so much verbally as by her recent actions, that she cannot fight any more, that she is giving up the effort. This defection seems to him profoundly unjust; even dishonourable. For years the Tates' domestic life has been governed according to the principle of separation of powers: Erica functioning as the executive branch, and Brian as the legislative and judicial. He has always left it to her to supervise the children in everyday matters.

Now when—possibly as a consequence of her management—the children have grown into selfish, rude, rebellious adolescents, she resigns and declares that it is his turn. Which is as if the President and his cabinet should abdicate and turn over the task of suppressing a colonial revolt to Congress and the Supreme Court.

ALISON LURIE, *The War Between the Tates*, 1974

What was clear to any spectator was that the only link binding her to either parent was this lamentable fact of her being a ready vessel for bitterness, a deep little porcelain cup in which biting acids could be mixed. They had wanted her not for any good they could do her, but for the harm they could, with her unconscious aid, do each other. She should serve their anger and seal their revenge,

HENRY JAMES, *What Maisie Knew*, 1897

FAR STAR

It is like living in a transistor with all this radio
Which is the inner weather of the house
Presided over by housegoddesses who turn
Everything that happens into perfume and electricity;
Oh! she cries, what a blessing—and I smell the blessing
Like a candle lighted; a scented flame that spreads
Through closed doors, opening them;
And when she curses, sulphur blackens all the knives.
We have tuned our circuits by living together so long
And the child, never having known another house, deepest tuned:
She was broadcast into this world via the lady transmitter
And mostly plays musical comedy, though now is of an age
For an occasional tragic aria about the sister she has not got,
Who will not now be broadcast from that far star;

And I wish heartily we had more loos—our tuning is such
On the same channel that we all three must shit simultaneously.

PETER REDGROVE, 1987

Only the really loving woman, the woman who is happier in giving than in taking, who is firmly rooted in her own existence, can be a loving mother when the child is in the process of separation.

Motherly love for the growing child, love which wants nothing for oneself, is perhaps the most difficult form of love to be achieved, and all the more deceptive because of the ease with which a mother can love her small infant. But just because of this difficulty, a woman can be a truly loving mother only if she can *love*; if she is able to love her husband, other children, strangers, all human beings. The woman who is not capable of love in this sense can be an affectionate mother as long as the child is small, but she cannot be a loving mother, the test of which is the willingness to bear separation—and even after the separation to go on loving.

ERICH FROMM, *The Art of Loving*, 1957

Your children are not your children
They are the sons and daughters of
Life's longing for itself.
They come through you but not from you
And though they are with you, yet they belong not to you.
You may give them your love but not your thoughts,
For they have their own thoughts.
You may house their bodies but not their souls,
For their souls dwell in the house of tomorrow which you
cannot visit, not even in your dreams.
You may strive to be like them, but seek not to make them like
you.
For life goes not backward nor tarries with yesterday.

KAHLIL GIBRAN, *The Prophet*, 1923

7

Dangerous Liaisons

'I CONSIDER a country-dance as an emblem of marriage. Fidelity and complaisance are the principal duties of both; and those men who do not choose to dance or marry themselves, have no business with the partners or wives of their neighbours', says Henry Tilney in *Northanger Abbey*.

An admirable precept, but William James was probably right when he had his revelation: man is basically polygamous, though in the modern world, where society exacts few penalties for adultery, women may be just as often tempted to infidelity as men.

Perhaps Strindberg is correct in regarding faithfulness as a characteristic rather than a virtue. Moreover, *autre pays, autres mœurs*: there are cultures where the mistress has a recognized position, and, as Sir James Goldsmith remarked, 'when you marry your mistress, you automatically create a vacancy'.

The exhilaration of falling in love anew can be hard to resist, and for some there is a constant search for that intense experience, often accompanied by the conviction that this time will be different, and will be the last. 'These illusions are greatly helped by the deceptive character of sexual desire' (Erich Fromm, *The Art of Loving*).

Cuckolds and unfaithful husbands are among the oldest butts in the world, and there are many lightheartedly bawdy accounts of adultery. But the mere suspicion of a partner's infidelity, even if unfounded, can be agonizing, as the letters of some of the greatest men in history, Mozart and Churchill among them, testify, and jealousy itself can corrode a relationship or even destroy a personality, as it did the hero of Trollope's *He Knew He Was Right*.

There are couples for whom periodic infidelities are a way of life. 'She saw life as a perpetual sex-battle between husbands who desire to be unfaithful to their wives, and wives who desire to recapture their husbands in the end' (Ford Madox Ford, *The Good Soldier*).

Some marriages may even remain untouched by occasional unfaithfulness, but usually there is a price to pay for such a diversion of energies from the marriage. 'Marriage drinks up all our power of giving or getting any blessedness in that sort of love. I know it may be very dear—but it murders our marriage—and then the marriage stays with us like a murder—and everything else is gone' (George Eliot, *Middlemarch*). Pepys's astonishingly honest diaries have charted the devastating effect on a couple of the discovery of infidelity: the wild see-saw of emotion from self-destructive jealousy and unforgiving rage to heightened intimacy.

A marriage may also be strengthened by such an extra-marital episode. It can force the couple to make a more realistic appraisal of their relationship, and repentance and forgiveness may form a coda: emotions often best expressed in music, and most sublimely in the Countess's aria at the end of Mozart's Figaro.

William James, psychologist and philosopher, woke one night feeling he had solved the ultimate mystery of life. The following morning he found that this doggerel was the great insight he had written down:

Hogamous, higamous
Man is polygamous
Higamous, hogamous
Woman monogamous

THE THIRD JUNGLE BOOK

Why does the Pygmy
Indulge in polygmy?
His tribal dogma
Frowns on monogma.

OGDEN NASH, 1943

Now things are so ill arranged that some people are born monogamous, that is, faithful, which is not a virtue but a quality, while others are born polygamous, that is, unfaithful. If these two opposites come together the result is great misery.

AUGUST STRINDBERG, Preface to *Getting Married*, 1884

HEAVEN

In the heaven of the god I hope for (call him X)
There is marriage and giving in marriage and transient sex
For those who will cast the body's vest aside
Soon, but are not yet wholly rarefied
And still embrace. For X is never annoyed
Or shocked; has read his Jung and knows his Freud,
He gives you time in heaven to do as you please,
To climb love's gradual ladder by slow degrees,
Gently to rise from sense to soul, to ascend
To a world of timeless joy, world without end.

A. S. J. TESSIMOND, 1978

You may meet in life (as in literature) women who are flighty, or even plain wanton—I don't refer to mere flirtatiousness, the sparring practice for the real combat, but to women who are too silly to take even love seriously, or are actually so depraved as to enjoy 'conquests', or even enjoy the giving of pain—but these are abnormalities, even though false teaching, bad upbringing, and corrupt fashions may encourage them. . . .But they are instinctively, when uncorrupt, monogamous. *Men are not*. . . . No good pretending. Men just ain't, not by their animal nature. Monogamy (although it has long been fundamental to our inherited *ideas*) is for us men a piece of 'revealed' ethic, according to faith and not to the flesh. Each of us could healthily beget, in our 30 odd years of full manhood, a few hundred children, and enjoy the process. . . .

Faithfulness in Christian marriage entails that: great mortification. For a Christian man there is *no escape*. Marriage may help to sanctify and direct to its proper object his sexual desires; its grace may help him in the struggle; but the struggle remains. It will not satisfy him—as hunger may be kept off by regular meals. It will offer as many difficulties to the purity proper to that state, as it provides easements. No man, however truly he loved his betrothed and bride as a young man, has lived faithful to her as a wife in mind and body without deliberate conscious exercise of the *will*, without self-denial.

J. R. R. TOLKIEN, from a letter to his son Michael, 6–8 March 1941

If the purpose of dinner is to nourish the body the man who eats two dinners at a sitting may perhaps attain greater enjoyment but not his object, since the stomach will not digest two dinners.

If the purpose of marriage is the family the person who seeks to have a number of wives or husbands may possibly obtain much pleasure therefrom, but will not in any case have a family.

If the purpose of food is nourishment and the purpose of marriage is the family the whole question resolves itself into not eating more than the stomach can digest and not having more wives or husbands than are needed for the family—that is, one wife or one husband. Natasha needed a husband. A husband was given her. And her husband gave her a family. And she not only saw no need of any other or better husband but as all her spiritual energies were devoted to serving that husband and family she could not imagine, and found no interest in imagining, how it would be if things were different.

LEO TOLSTOY, *War and Peace*, 1863–9

LEONTES: Ha' not you seen, Camillo?
(But that's past doubt; you have, or your eye-glass
Is thicker than a cuckold's horn,) or heard?
(For to a vision so apparent rumour
Cannot be mute) or thought? (for cogitation
Resides not in that man that does not think)
My wife is slippery? If thou wilt confess,
Or else be impudently negative,
To have nor eyes, nor ears, nor thought, then say
My wife's a hobby-horse, deserves a name
As rank as any flax-wench that puts to
Before her troth-plight: say't, and justify't.
CAMILLO: I would not be a stander-by to hear
My sovereign mistress clouded so, without
My present vengeance taken: 'shrew my heart,
You never spoke what did become you less
Than this; which to reiterate were sin
As deep as that, though true.
LEONTES: Is whispering nothing?
Is leaning cheek to cheek? is meeting noses?
Kissing with inside lip? stopping the career
Of laughter with a sigh (a note infallible
Of breaking honesty)? horsing foot on foot?
Skulking in corners? wishing clocks more swift?
Hours, minutes? noon, midnight? and all eyes
Blind with the pin and web but theirs; theirs only,
That would unseen be wicked? is this nothing?
Why, then the world, and all that's in't, is nothing,
The covering sky is nothing, Bohemia nothing,
My wife is nothing, nor nothing have these nothings,
If this be nothing.
CAMILLO: Good my lord, be cur'd
Of this diseas'd opinion, and betimes;
For 'tis most dangerous.

WILLIAM SHAKESPEARE, *The Winter's Tale*, 1661

Dear little wife! I want to talk to you quite frankly. You have no reason whatever to be unhappy. You have a husband who loves you and does all he possibly can for you. As for your foot, you must just be patient and it will surely get well again. I am glad indeed when you have some fun—of course I am—but I do wish that you would not sometimes make yourself so cheap. In my opinion you are too free and easy with N.N. . . . and it was the same with N.N., when he was still at Baden. Now please remember that N.N. are not half so familiar with other women, whom they perhaps know more intimately, as they are with you. Why, N.N. who is usually a well-conducted fellow and particularly respectful to women, must have been misled by your behaviour into writing the most disgusting and most impertinent sottises which he put into his letter. A woman must always make herself respected, or else people will begin to talk about her. My love! Forgive me for being so frank, but my peace of mind demands it as well as our mutual happiness. Remember that you yourself once admitted to me that you were inclined to *comply too easily*. You know the consequences of that. Remember too the promise you gave to me. Oh, God, do try, my love!

WOLFGANG AMADEUS MOZART, from a letter to his wife, August 1789

OTHELLO: I had been happy, if the general camp,
Pioners, and all, had tasted her sweet body,
So I had nothing known. O, now for ever
Farewell the tranquil mind, farewell content;
Farewell the plumed troop, and the big wars
That make ambition virtue! O, farewell,
Farewell the neighing steed, and the shrill trump,
The spirit-stirring drum, the ear-piercing fife,
The royal banner, and all quality,
Pride, pomp, and circumstance of glorious war!
And, O ye mortal engines, whose wide throats
The immortal Jove's great clamour counterfeit,
Farewell! Othello's occupation's gone!

WILLIAM SHAKESPEARE, *Othello*, 1604

To citizen Bonaparte,
care of citizen Beauharnais,
6, rue Chantereine,
Paris.

Nice, 10 Germinal, year IV [1796]

I have not spent a day without loving you; I have not spent a night without embracing you; I have not so much as drunk a single cup of tea without cursing the pride and ambition which force me to remain separated from the moving spirit of my life. In the midst of my duties, whether I am at the head of my army or inspecting the camps, my beloved Josephine stands alone in my heart, occupies my mind, fills my thoughts. If I am moving away from you with the speed of the Rhône torrent, it is only that I may see you again more quickly. If I rise to work in the middle of the night, it is because this may hasten by a matter of days the arrival of my sweet love. Yet in your letter of the 23rd. and 26th. Ventôse, you call me *vous*. *Vous* yourself! Ah! wretch, how could you have written this letter? How cold it is! And then there are those four days between the 23rd. and the 26th.; what were you doing that you failed to write to your husband? . . . Ah, my love, that *vous*, those four days make me long for my former indifference. Woe to the person responsible! May he, as punishment and penalty, experience what my convictions and the evidence (which is in your friend's favour) would make me experience! Hell has no torments great enough! Nor do the Furies have serpents enough! *Vous! Vous!* Ah! how will things stand in two weeks? . . . My spirit is heavy; my heart is fettered and I am terrified by my fantasies. . . . You love me less; but you will get over the loss. One day you will love me no longer; at least tell me; then I shall know how I have come to deserve this misfortune. . . . Farewell, my wife: the torment, joy, hope and moving spirit of my life; whom I love, whom I fear, who fills me with tender feelings which draw me close to Nature, and with violent impulses as tumultuous as thunder. I ask of you neither eternal love, nor fidelity, but simply . . . *truth*, unlimited honesty. The day when you say 'I love you less', will mark the end of my love and the last day of my life. If my heart were base enough to love without being loved in return I would tear it to pieces. Josephine! Josephine! Remember what I have sometimes said to you: Nature has endowed me with a virile and decisive character. It has built yours out of lace and gossamer. Have you ceased to love me?

Forgive me, love of my life, my soul is racked by conflicting forces.

My heart, obsessed by you, is full of fears which prostrate me with misery. . . . I am distressed not to be calling you by name. I shall wait for you to write it.

Farewell! Ah! if you love me less you can never have loved me. In that case I shall truly be pitiable.

NAPOLEON BONAPARTE, letter to his wife

Dearest, it worries me vy much that you should seem to nurse such absolutely wild suspicions wh are so dishonouring to all the love & loyalty I bear you & will please god bear you while I breathe. They are unworthy of you & me. And they fill my mind with feelings of embarrassment to wh I have been a stranger since I was a schoolboy. I know that they originate in the fond love you have for me, and therefore they make me feel tenderly towards you & anxious always to deserve that most precious possession of my life. But at the same time they depress me & vex me—& without reason.

We do not live in a world of small intrigues, but of serious & important affairs. I could not conceive myself forming any other attachment than that to which I have fastened the happiness of my life here below.

And it offends my best nature that you should—against your true instinct—indulge small emotions & wounding doubts. You ought to trust me for I do not love & will never love any woman in the world but you and my chief desire is to link myself to you week by week by bonds which shall ever become more intimate & profound.

Beloved I kiss your memory—your sweetness & beauty have cast a glory upon my life.

You will find me always your loving & devoted husband

W

WINSTON S. CHURCHILL, letter to his wife, November 1909

ANGELO: 'Tis one thing to be tempted, Escalus,
Another thing to fall.

WILLIAM SHAKESPEARE, *Measure for Measure*, 1604

A LOYAL WIFE

My lord, I am grateful for these two pearls you offer me. I tremble with uncertainty. What shall I say? I say to you . . . I am married and have sworn to be faithful to my husband.

Perhaps you do not know that the colors of my family hang in the Royal Park? Perhaps you do not know that my husband is honorary lancer in the Emperor's Palace?

I think you are sincere; I think you are honorable. Therefore I have put your pearls against my robe, and looked at them, and smiled. But take them now again. Perhaps you will take these two tears as well?

TCHANG TSI, *c.* AD 800

Let us suppose that you do really love me (and it is only so as to avoid returning in future to this subject that I allow such a supposition) would the obstacles that lie between us be any the less insurmountable? And could I even then do anything else than wish you might soon overcome your love; above all, by hastening to deprive you of all hope, do everything in my power to help you overcome it? You yourself admit that love is 'a painful feeling when it is not shared by the one who inspires it'. Now, you know quite well that it is impossible for me to share it; even were such a misfortune to befall me, I should suffer for it while you yourself would be none the happier. I hope you respect me enough not to doubt that for an instant. No more then, I beg you; give up your attempts to disturb a heart which is so much in need of peace. Do not oblige me to regret having known you.

Cherished and esteemed by a husband whom I love and respect, my duties and my pleasures find in him their common source. I am happy: I have a right to be so. If livelier pleasures exist, I have no desire for them: I do not wish to know them. Can any pleasure be sweeter than that of living at peace with oneself, passing one's days in serenity, sleeping untroubled, waking without remorse? What you call happiness is nothing but a tumult in the mind, a tempest of passion, frightful to behold even for the spectator on the shore. Come, how could I face such storms? How dare to embark upon a sea strewn with so many thousand wrecks? And with whom? No,

Monsieur, I shall hold my ground: I cherish the ties that keep me there. I would not break them, if I could, and if I had none I should hasten to make them.

Why do you dog my footsteps? Why do you so obstinately pursue me? Your letters, which were to have come seldom, arrive in rapid succession. They were to have been discreet and they speak of nothing but your insane love. I am more beset by the thought of you than I ever was by your person. Repulsed under one guise you reappear in another. Things you are asked not to say you say again, but in a different way. You are ready to entangle me in specious arguments, but refuse to answer mine. I do not wish to reply to any more of your letters. I shall not reply to them. . . . How you treat the women you have seduced! With what contempt you speak of them! I can believe that some of them deserve it, but were they all so despicable? Ah, of course—since they betrayed their duty in surrendering to unlawful love. At that moment they lost everything, even the respect of the man for whom they made their sacrifice. The punishment is just, but the mere thought of it makes one shudder. What is it to me, after all? Why should I care about them or about you? What right have you to disturb my peace of mind? Leave me alone, see me no more, write to me no more, I beg of you; I insist. This letter is the last you will receive from me.

PIERRE CHODERLOS DE LACLOS, *Les Liaisons Dangereuses*, 1782

To be the county family, to look the county family, to be so appropriately and perfectly wealthy; to be so perfect in manner—even just to the saving touch of insolence that seems to be necessary. To have all that and to be all that! No, it was too good to be true. And yet, only this afternoon, talking over the whole matter she said to me: 'Once I tried to have a lover but I was so sick at the heart, so utterly worn out that I had to send him away.' That struck me as the most amazing thing I had ever heard. She said 'I was actually in a man's arms. Such a nice chap! Such a dear fellow! And I was saying to myself, fiercely, hissing it between my teeth, as they say in novels—and really clenching them together: I was saying to myself: "Now, I'm in for it and I'll really have a good time for once in my life—for once in my life!" It was in the dark, in a carriage, coming back from a hunt-ball. Eleven miles we had to drive! And then suddenly the bitterness of the endless poverty, of the endless acting—it fell on me like a blight, it spoilt everything. Yes, I had to realize that I had been spoilt even for the good time when it came. And I burst out crying and I cried and I cried for the whole eleven miles. Just imagine *me* crying! And just imagine me making a fool of the poor dear chap like that. It certainly wasn't playing the game, was it now?'

I don't know; I don't know; was that last remark of hers the remark of a harlot, or is it what every decent woman, county family or not county family, thinks at the bottom of her heart? Or thinks all the time for the matter of that? Who knows?

FORD MADOX FORD, *The Good Soldier*, 1915

That Saturday morning at the cottage John was woken by the sound of his children quarrelling amicably in the kitchen below. He saw from the clock on his bedside table that it was eight o'clock. Tom and Anne, it seemed, were getting their own breakfast so John turned over in bed and closed his eyes again.

He could not sleep. He could rarely sleep again once he was awake, but because it was Saturday he felt some sort of obligation to lie in, as if the extra hour was a luxury he must enjoy like sun-bathing or stereophonic gramophone music. As a result his mind filled up neither with the nonsensical and irrational dreams which came to him when he was truly unconscious, nor with the incisive, systematized thoughts which when he was awake made him so good at the law: instead vague anxieties and fantasies flitted in and out of his mind, settling promiscuously on such subjects as income tax, the garden or a new car.

On this particular Saturday morning his semi-conscious mind went back over the week in London and settled—half-anxiously, half-pleasurably—on the thought of Jilly Mascall. And having remembered the past week it moved forward to the future: in anticipation of their new assignation he began to imagine how it would be. He smiled to himself as he saw her laugh at his amusing conversation: she directs more of those straight, sultry looks at him—and towards the end of their lunch she complains that the drain in her kitchen sink is blocked, and that although she has a plunger neither she nor Miranda is strong enough to dislodge the nugget of tea leaves and bacon fat or whatever it is . . .

John has nothing on that afternoon so he offers to return with her and see what he can do. He goes back to the flat. He clears the drain.

She offers to make some coffee. They sit sipping, talking. He moves towards her. They kiss. Her passive, inexperienced lips press tentatively against his. He holds her. He touches her. Patiently, kindly, he removes her clothes. Desire and gratitude mingle in her eyes. He uncovers the pink, pigmented breasts with their tiny, child-like nipples. His hand stretches down to her stomach. He hesitates, remembering how young she is. 'Oh please,' she murmurs. They go ahead. With the care of an artist and the skill of a surgeon he makes love to her until, with an ecstatic whimper, then a cry, from her pink little mouth, she feels for the first time the consummation of sexual love.

Next to him in bed Clare shifted and sighed. There was a muffled fart under the blankets. She raised herself on her elbow and leaned over her husband to look at the clock. The smell of her wind mingled with the liverish, early-morning odour from her mouth. She slumped back on her side of the bed, grunted, heaved around like an elephant and then was still. It was not yet time to get up.

In that his body had followed his mind so far in his imaginings but not all the way, John considered a move towards the passive haunch of his semi-slumbering wife: but since the contrast between the pink purity of Jilly Mascall and the heavy, smelly Clare Strickland was so great; and since the children might burst in upon them at any moment; and since his need to urinate was greater than his need to ejaculate, John rolled out of bed instead and stumbled through to the bathroom.

PIERS PAUL READ, *A Married Man*, 1979

IN MY FASHION

Dear, they said that woman resembled you.
Was that why I went with her, flirted with her,
raised my right hand to her left breast
till I heard the still sad music of humanity?
I complimented you! Why do you object?

Still you shrill, discover everything untrue:
your doppelgänger does not own your birthmarks,
cannot know our blurred nights together.
That music was cheap—a tune on a comb at best,
harsh and grating. Yes, you chasten me

and subdue. Well, that woman was contraband
and compared with you mere counterfeit.
Snow on the apple tree is not apple blossom—
all her colours wrong, approximate,
as in a reproduction of a masterpiece.

DANNIE ABSE, 1981

THE UNFAITHFUL WIFE

So I took her to the riverside
taking her for a virgin
but she had taken a husband.
Midsummer night, the Feast of St James:
it was a point of honour.
The streetlamps went off
and the crickets went on.
On the outskirts
I touched her sleeping breasts
and instantly they blossomed
like sprays of hyacinth.
The starch in her underskirt
grated on my ears like
a sheet of silk
lacerated by ten knives.
Without silver glinting on their leaves
the trees looked massive.
A horizon of dogs
Howls in the distance.

Past the blackberries,
the rushes, the hawthorn,
under her hair in the sand
I made a hollow for her head.
I took off my tie,
she took off her dress.
I removed my gunbelt,
she removed her underwear.
Neither petals nor shells
match such delicate skin,
nor do moonlit mirrors

glow with such shine.
Her thighs struggled
like astonished fish
caught in a torrent
now fiery, now frozen.
That night I rode
on the best of roads,
my mother-of-pearl mount
free and unbridled.
As a man I hesitate to say
the things she said to me:
the seeds of understanding
breed discretion.
Covered in kisses and sand
I removed her from the riverside.
The swords of the lillies
cut through the atmosphere.

I acted like the man I am,
a decent gypsy.
I gave her a big sewing-basket
of straw-coloured satin,
and preferred not to fall in love
because she had taken a husband
yet told me she was a virgin
when I took her by the riverside.

FREDERICO GARCÍA LORCA (1898–1936)

MAY I FEEL SAID HE

may i feel said he
(i'll squeal said she
just once said he)
it's fun said she

(may i touch said he
how much said she
a lot said he)
why not said she

(let's go said he
not too far said she
what's too far said he
where you are said she)

may i stay said he
(which way said she
like this said he
if you kiss said she

may i move said he
is it love said she)
if you're willing said he
(but you're killing said she

but it's life said he
but your wife said she
now said he)
ow said she

(tiptop said he
don't stop said she
oh no said he)
go slow said she

(cccome? said he
ummm said she)
you're divine! said he
(you are Mine said she)

E. E. CUMMINGS, 1935

OF AN HEROICAL ANSWER OF A GREAT ROMAN LADY TO HER HUSBAND

A grave wise man that had a great rich lady,
Such as perhaps in these days found there may be,
Did think she played him false and more than think,
Save that in wisdom he thereat did wink.

Howbeit one time disposed to sport and play
Thus to his wife he pleasantly did say,
'Since strangers lodge their arrows in thy quiver,
Dear dame, I pray you yet the cause deliver,
If you can tell the cause and not dissemble,
How all our children me so much resemble?'
The lady blushed but yet this answer made
'Though I have used some traffic in the trade,
And must confess, as you have touched before,
My bark was sometimes steered with foreign oar,
  Yet stowed I no man's stuff but first persuaded
  The bottom with your ballast full was laded.'

SIR JOHN HARINGTON (*c*.1561–1612)

¶ *Sir John Harington was a godson of Elizabeth I, who said to him when on her deathbed: 'When thou dost feel creeping time at thy gate, these fooleries will please thee less.'*

  Fresh-hearted May on hearing what he said
Benignly answered him with drooping head,
But first and foremost she began to weep.
'Indeed,' she said, 'I have a soul to keep
No less than you, and then there is my honour
Which for a wife is like a flower upon her.
I put it in your hands for good or ill
When the priest bound my body to your will,
So let me answer of my own accord
If you will give me leave, beloved lord;
I pray to God that never dawn the day
—Or let me die as foully as I may—
When I shall do my family that shame
Or bring so much dishonour on my name
As to be false. And if my love grow slack,
Take me and strip me, sew me in a sack
And drop me in the nearest lake to drown.
I am no common woman of the town,
I am of gentle birth, I keep aloof.
So why speak thus to me, for what reproof

Have I deserved? It's men that are untrue
And women, women ever blamed anew.
I think it a pretence that men profess;
They hide behind a charge of faithlessness.'
  And as she spoke she saw a short way off
Young Damian in his bush. She gave a cough
And signalled with a finger quickly where
He was to climb into a tree—a pear—
Heavily charged with fruit, and up he went,
Perfectly understanding what she meant,
Or any other signal, I may state,
Better than January could, her mate.
For she had written to him, never doubt it,
Telling him all and how to set about it.
And there I leave him sitting, by your pardon,
While May and January roamed the garden.

*

So long among the paths had strayed their feet
That they at last had reached the very tree
Where Damian sat in waiting merrily,
High in his leafy bower of fresh green.
And fresh young May, so shiningly serene,
Began to sigh and said 'Oh! I've a pain!
Oh Sir! Whatever happens, let me gain
One of those pears up there that I can see,
Or I shall die! I long so terribly
To eat a little pear, it looks so green.
O help me for the love of Heaven's Queen!
I warn you that a woman in my plight
May often feel so great an appetite
For fruit that she may die to go without.'
  'Alas,' he said, 'that there's no boy about,
Able to climb. Alas, alas,' said he,
'That I am blind.' 'No matter, sir,' said she,
'For if you would consent—there's nothing in it—
To hold the pear-tree in your arms a minute
(I know you have no confidence in me),
Then I could climb up well enough,' said she,
'If I could set my foot upon your back.'

'Of course,' he said, 'why, you shall never lack
For that, or my heart's blood to do you good.'
And down he stooped; upon his back she stood,
Catching a branch, and with a spring she thence
—Ladies, I beg you not to take offence,
I can't embellish, I'm a simple man—
Went up into the tree, and Damian
Pulled up her smock at once and in he thrust.
And when King Pluto saw this shameful lust
He gave back sight to January once more
And made him see far better than before.
Never was man more taken with delight
Than January when he received his sight.
And his first thought was to behold his love.
He cast his eyes into the tree above
Only to see that Damian had addressed
His wife in ways that cannot be expressed
Unless I use a most discourteous word.
He gave a roaring cry, as might be heard
From stricken mothers when their babies die.
'Help! Out upon you!' He began to cry.
'Strong Madam Strumpet! What are you up to there?'
'What ails you, sir?' said she, 'what makes you swear?
Have patience, use the reason in your mind,
I've helped you back to sight when you were blind!
Upon my soul I'm telling you no lies;
They told me if I wished to heal your eyes
Nothing could cure them better than for me
To struggle with a fellow in a tree.
God knows it was a kindness that I meant.'
'Struggle?' said he, 'Yes! Anyhow, in it went!
God send you both a shameful death to die!
He had you, I saw it with my very eye,
And if I did not, hang me by the neck!'
'Why then,' she said, 'my medicine's gone to wreck,
For certainly if you could really see
You'd never say such words as those to me;
You caught some glimpses, but your sight's not good.'
'I see,' he said, 'as well as ever I could,
Thanks be to God! And with both eyes, I do!

And that, I swear, is what he seemed to do.'
'You're hazy, hazy, my good sir,' said she;
'That's all I get for helping you to see.
Alas,' she said, 'that ever I was so kind!'
'Dear wife,' said January, 'never mind,
Come down, dear heart, and if I've slandered you
God knows I'm punished for it. Come down, do!
But by my father's soul, it seemed to me
That Damian had enjoyed you in the tree
And that your smock was pulled up over your breast.'
'Well, think,' she said, 'as it may please you best,
But, Sir, when suddenly a man awakes,
He cannot grasp a thing at once, it takes
A little time to do so perfectly,
For he is dazed at first and cannot see.
Just so a man who has been blind for long
Cannot expect his sight to be so strong
At first, or see as well as those may do
Who've had their eyesight back a day or two.
Until your sight has settled down a bit
You may be frequently deceived by it.
Be careful then, for by our heavenly King
Many a man feels sure he's seen a thing
Which was quite different really, he may fudge it;
Misapprehend a thing and you'll misjudge it.'
And on the word she jumped down from the tree.
And January—who is glad but he?—
Kissed her and clasped her in his arms—how often!—
And stroked her womb caressingly to soften
Her indignation. To his palace then
He led her home. Be happy, gentlemen,
That finishes my tale of January;
God and his Mother guard us, blessed Mary!

GEOFFREY CHAUCER, *The Merchant's Tale*, c.1390

THE 'WOOING' OF MR MONROE

Little Mrs Monroe met the challenge of the very blonde lady with all of her charming directness. She went to Miss Lurell's apartment and said to her, quite simply, 'I am Mrs John Monroe. I have come to tell you some things about John I think you should know.' The other woman met her simplicity with icy reticence. 'Please understand,' pursued Mrs Monroe, 'that I do not wish to interfere. John has told me of the strange beauty of it all. I just wanted to warn you that John is simply terrible with machinery.'

'There is no machinery in our association that I can think of,' said the lovely Miss Lurell, coldly.

'Oh,' said Mrs Monroe, 'there will be. May I smoke?' She lighted a cigarette, her first in months. 'Machinery is always bobbing up in John's life. He knows nothing at all about it, but I will say for him that he never runs from it. I might almost say that he attacks it. He attacks machinery.'

'I don't believe I understand,' said the other, as if to imply that she did not wish to understand. Mrs Monroe was about to inhale some smoke, thought better of it—she always choked—and smiled amicably.

'Not long ago,' she began, 'we went for a motor trip to John's university; he hadn't been back there for years. We stayed at a charming place on the campus, called the Union. It was very peaceful. We could see apple trees in blossom, from our window. It was early May—'

'Pray spare yourself memories which can only hurt,' murmured the other woman.

'Oh, it was really quite funny.' Mrs Monroe permitted herself what she had intended to be a gay, rippling laugh. 'We had been in our room only ten minutes, when John went across the hall to take a shower in a great tiled shower-room for alumni guests. I remember it was twilight, soft and dreamy—' Miss Lurell made a sound as of one who dreads sentimental tears. 'Well,' continued Mrs Monroe, 'John had forgotten to bring his bathrobe, of course, so he wore his raincoat. He always forgets his bathrobe—and theatre tickets.'

'I don't see what you can possibly hope to establish—by all this,' interrupted the blonde lady.

'I am telling you such an intimate story, because this was so typical of John,' said his wife. 'You see he has never really taken a successful shower in his life. He always gets the water to running too cold or too hot. This time it ran too cold. He kept twisting the handle and swearing until a man in an adjoining shower told him to turn it farther to the right. John shot it all the way to the right. Instantly a stream of boiling water flooded the bath. John didn't get scalded, because he has learned not to get fully into a shower: he stands outside and sticks his feet in and then his shoulders. You see, I knew you wouldn't have had any experience of John's showers—'

'You were quite right,' said the other, frigidly.

'Well, in a few seconds the whole place was a fog of steam and the heat was frightful. John couldn't reach into the compartment again and turn the handle back, so he began to go "Woo! Woo!"—like a child. He always goes "Woo! Woo!" when things go wrong with machinery. Of course he writes beautiful sonnets, which I am sure you appreciate perhaps more deeply than I do, and of course mechanical things are of no importance, but one must know what to do with him in a case like this.'

'And what did *you* do?' asked Miss Lurell.

'Well, my dear, first the other man in the boiler-room—as it had now become—climbed up on the wall of his shower and tried to reach over and get at the handle from above, but the intense heat made that impossible. Then he yelled at John to get a window-stick or something with which he could reach down and knock the handle back onto "Cold." Of course John was too excited to be of any use himself. Finally, in a panic, he rushed across the hall into my room, stark, raving—'

'Please!' said Miss Lurell.

'Stark, raving naked,' continued Mrs Monroe. 'He's so funny that way, really I just *screamed*. He was still making that "Woo! Woo!" noise and I knew instantly he had been fooling with the works of something. It was just the way he acted the time he short-circuited all the lights in a theatre one night between acts—we never found out how he did that—he got to wandering around and stumbled into a switch or something, probably thinking it was a water-cooler.'

'Fully dressed, I presume?'

'Oh, it's only when he's driven from a shower or something like that that he hasn't anything on,' said Mrs Monroe, simply. 'Well, he began yelling at me to get him a window-stick or something and finally pulled down a curtain rod, curtain and all, and would have rushed back across the hall with it the way he was, but I threw my negligée over him. When he got back the other man had had to leave, for the heat was unbearable. In the end, the university engineer had to shut off the water in the whole institution—all the campus buildings—they phoned him to do that because there was nobody in the

Union who knew where the local water-switch was—I mean the one for that building. It was terrible, but that's the way John is—when he fools with machinery, he always disconnects the whole works. Once in a hotel in Nice, he—'

'May I ask,' cut in the other woman, 'how long you have endured this?'

'Eight years in June,' said Mrs Monroe. 'Naturally, I feel that the—next lady—should know what to expect.'

'Eight years,' murmured Miss Lurell. She rose. Mrs Monroe rose too.

'*Now* you will know what to do,' said Mrs Monroe. 'Don't argue with him when he begins to "Woo!"—just let him have his own way, but summon somebody instantly.'

'I know exactly what to do,' said Miss Lurell, with an odd smile. She accompanied little Mrs Monroe to the door, where she impulsively held out her hand. 'Apropos of nothing,' drawled the very blonde lady, 'may I ask if you play bridge?'

'Oh, very badly,' said Mrs Monroe. 'Unless—' she waved a gloved hand at a passing taxi, 'unless I hold a perfect grand slam.' She smiled back, over her shoulder, and went away.

JAMES THURBER, *The Owl in the Attic and Other Perplexities*, 1931

Everything had gone wrong in the Oblonsky household. The wife had found out about her husband's relationship with their former French governess and had announced that she could not go on living in the same house with him. This state of affairs had already continued for three days and was having a distressing effect on the couple themselves, on all the members of the family, and on the domestics. They all felt that there was no sense in their living together under the same roof and that any group of people who chanced to meet at a wayside inn would have more in common than they, the members of the Oblonsky family, and their servants. The wife did not leave her own rooms and the husband stayed away from home all day. The children strayed all over the house, not knowing what to do with themselves. The English governess had quarrelled with the housekeeper and had written a note asking a friend to find her a new place. The head-cook had gone out right at dinner-time the day before. The under-cook and the coachman had given notice.

On the third morning after the quarrel Prince Stepan Arkadyevich

Oblonsky—Stiva, as he was generally called by his friends—awoke at his usual time, which was about eight o'clock, not in his wife's bedroom but on a morocco-leather couch in his study. He turned his plump, pampered body over on the springs, as if he had a mind for a long sleep, and hugged the pillow, pressing his cheek to it; but with a start he sat up on the sofa and opened his eyes.

*

'Oh dear, dear, dear!' he groaned, remembering what had happened. And he went over all the details of the scene with his wife, seeing the complete hopelessness of his position and, most tormenting thought of all, the fact that it was his own fault.

'No, she will never forgive me—she cannot forgive me! And the worst of it is that I am to blame for everything—I am to blame and yet I am not to blame. That is the whole tragedy,' he mused, and recalled despairingly the most painful aspects of the quarrel.

That first moment had been the worst, after he had returned from the theatre, happy and content, with a huge pear in his hand for his wife, and had not found her in the drawing-room. Nor, to his astonishment, was she in the study. At last he had discovered her in the bedroom, holding the wretched note that explained everything.

Dolly, whom he always thought of as busy, bustling, and rather foolish, now sat motionless, looking at him with an expression of horror, despair, and anger, the note in her hand.

'What is this? What does this mean?' she had demanded, pointing to the note.

As the scene came back to him, as often happens it was not so much the memory of the event that tormented him, as of the way he had replied to her.

It had been with him as it so often is when people are unexpectedly caught out in something disgraceful. He had been unable to assume an expression suitable to the situation in which he was placed by his wife's knowledge of his guilt. Instead of taking offence or denying the whole thing, instead of justifying himself or begging forgiveness or even remaining indifferent—any of which would have been better than what he actually did—in spite of himself ('by a reflex action of the brain,' now thought Oblonsky, who had a leaning towards physiology), in spite of himself he suddenly smiled his habitual, kind, and somewhat foolish smile.

This foolish smile he could not forgive himself. Seeing it, Dolly had shuddered as if with a physical pain and bursting with her usual vehemence into a stream of bitter words she had fled from the room. Since then she had refused to see her husband.

'It's all because of that silly smile,' thought Oblonsky.

'But what is to be done? What can I do?' he asked himself despairingly and found no answer.

LEO TOLSTOY, *Anna Karenina*, 1873–7

What constitutes adultery is not the hour which a woman gives her lover, but the night which she afterwards spends with her husband.

GEORGE SAND (1804–76)

WITH HER LIPS ONLY

This honest wife, challenged at dusk
At the garden gate, under a moon perhaps,
In scent of honeysuckle, dared to deny
Love to an urgent lover: with her lips only,
Not with her heart. It was no assignation;
Taken aback, what could she say else?
For the children's sake, the lie was venial;
'For the children's sake', she argued with her conscience.
Yet a mortal lie must follow before dawn:
Challenged as usual in her own bed,
She protests love to an urgent husband,
Not with her heart but with her lips only;
'For the children's sake', she argues with her conscience,
'For the children'—turning suddenly cold towards them.

ROBERT GRAVES, 1953

Samuel Pepys's touchingly honest account of his brief affair with Deb Willet and his struggle to give up the object of his fantasies, illustrates the continuing reverberations in a marriage of even such a minor incident of infidelity, and especially the outbreaks of injured rage on the one hand, of penitence as well as defiance and backsliding on the other, alternating with moments of heightened intimacy between husband and wife. He developed a curious lingua franca*—based on*

Spanish, with a mixture of other languages, to disguise the more erotic passages.

27 September 1667. . . . While I was busy at the office, my wife sends for me to come to home, and what was it but to see the pretty girl [Deb Willet] which she is taking to wait upon her; and though she seems not altogether so great a beauty as she had before told me, yet endeed she is mighty pretty; and so pretty, that I find I shall be too much pleased with it, and therefore could be contented as to my judgment, though not to my passion, that she might not come, lest I may be found too much minding her, to the discontent of my wife. She is to come next week. . . . To the office again, my [mind] running on this pretty girl.

31 March 1668. I called Deb to take pen, ink, and paper and write down what things came into my head for my wife to do, in order to her going into the country; and the girl writing not so well as she would do, cried, and her mistress construed it to be sullenness and so was angry, and I seemed angry with her too; but going to bed, she undressed me, and there I did give her good advice and beso la, ella weeping still; and yo did take her, the first time in my life, sobra mi genu and did poner mi mano sub her jupes and toca su thigh, which did hazer me great pleasure; and so did no more, but besando-la went to my bed.

25 October. Lords Day. . . . after supper, to have my head combed by Deb, which occasioned the greatest sorrow to me that ever I knew in this world; for my wife, coming up suddenly, did find me imbracing the girl con my hand sub su coats; and endeed, I was with my main in her cunny. I was at a wonderful loss upon it, and the girl also; and I endeavoured to put it off, but my wife was struck mute and grew angry, and as her voice came to her, grew quite out of order; and I do say little, but to bed; and my wife said little also, but could not sleep all night; but about 2 in the morning waked me and cried, and fell to tell me as a great secret that she was a Roman Catholique and had received the Holy Sacrament; which troubled me but I took no notice of it, but she went on from one thing to another, till at last it appeared plainly her trouble was at what she saw; but yet I did not know how much she saw and therefore said nothing to her. But after her much crying and reproaching me with inconstancy and preferring a sorry girl before her, I did give her no provocations but did promise all fair usage to her, and love, and foreswore any hurt that I did with her—till at last she seemed to be at ease again.

26 October. And so toward morning, a little sleep; and so I, with some little repose and rest, rose, and up and by water to Whitehall, but with my mind mightily troubled for the poor girl, whom I fear I have undone by this, my [wife] telling me that she would turn her out of door. . . . Thence by coach home and to dinner, finding my wife mightily discontented and the girl sad, and no words from my wife to her. . . . I all the evening busy and my wife full of trouble in her looks; and anon to bed—where about midnight, she wakes me and there falls foul on me again, affirming that she saw me hug and kiss the girl; the latter I denied, and truly; the other I confessed and no more. . . . but did promise her perticular demonstrations of my true love to her, owning some indiscretion in what I did, but that there was no harm in it. She at last on these promises was quiet, and very kind we were, and so to sleep.

1 November. Lords Day. And so to supper and to bed—my mind yet at disquiet that I cannot be informed how poor Deb stands with her mistress, but I fear she will put her away; and the truth is, though it be much against my mind and to my trouble, yet I think it will be fit that she be gone, for my wife's peace and mine; for she cannot but be offended at the sight of her, my wife having conceived this jealousy of me with reason. And therefore, for that, and other reasons of expense, it will be best for me to let her go—but I shall love and pity her.

3 November. I observed my wife to eye my eyes whether I did ever look upon Deb; which I could not, but do now and then (and to my grief did see the poor wretch look on me and see me look on her, and then let drop a tear or two; which doth make my heart relent at this minute that I am writing this, with great trouble of mind, for she is endeed my sacrifice, poor girl); and my wife did tell me in bed, by the by, of my looking on other people, and that the only way is to put things out of sight.

6 November. Up, and presently my wife up with me, which she professedly now doth every day to dress me, that I may not see

Willet; and doth eye me whether I cast my eye upon her or no. And doth keep me from going into the room where she is among the upholsters at work in our blue chamber.

8 November. Lords Day. Up, and at my chamber all the morning, setting papers to rights with my boy. And so to dinner at noon, the girl with us; but my wife troubled thereat to see her, and doth tell me so; which troubles me, for I love the girl. . . . I could wish in that respect that she was out of the house, for our peace is broke to all of us while she is here. And so to bed—where my wife mighty unquiet all night, so as my bed is become burdensome to me.

9 November. Up, and I did by a little note which I flung to Deb, advise her that I did continue to deny that ever I kissed her, and so she might govern herself. The truth [is], that I did adventure upon God's pardoning me this lie, knowing how heavy a thing it would be for me to be the ruin of the poor girl; and next, knowing that if my wife should know all, it were impossible ever for her to be at peace with me again—and so our whole lives would be uncomfortable.

10 November. Up, and my wife still every day as ill as she is all night; will rise to see me outdoors, telling me plainly that she dares not let me see the girl; and so I out to the office, where all the morning; and so home to dinner, where I find my wife mightily troubled again, more then ever, and she tells me that it is from her examining the girl and getting a confession now from her of all, even to the very tocando su thing with my hand—which doth mightily trouble me, as not being able to foresee the consequences of it as to our future peace together. So my wife would not go down to dinner, reproaching me with my unkindness and perjury, I having denied my ever kissing her—as also with all her old kindnesses to me, and my ill-using of her from the beginning, and the many temptations she hath refused out of faithfulness to me; . . . all which I did acknowledge and was troubled for, and wept; and at last pretty good friends again, and so I to my office and there late, and so home to supper with her; and so to bed, where after half-an-hour's slumber, she wakes me and cries out that she should never sleep more, and so kept raving till past midnight, that made me cry and weep heartily all the while for her, and troubled for what she reproached me with as before; and at last, with new vows, and perticularly that I would myself bid the girl be gone and show my dislike to her—which I shall endeavour to perform, but with much trouble.

11 November. To supper and to bed; where after lying a little while, my wife starts up, and with expressions of affright and madness, as one frantic, would rise; and I would not let her, but burst out in tears myself; and so continued almost half the night, the moon shining so that it was light; and after much sorrow and reproaches and little ravings (though I am apt to think they were counterfeit from her), and my promise again to discharge the girl myself, all was quiet again and so to sleep.

12 November. I to my wife and to sit with her a little; and then called her and Willet to my chamber, and there did with tears in my eyes, which I could not help, discharge her and advise her to be gone as soon as she could, and never to see me or let me see her more while she was in the house; which she took with tears too, but I believe understands me to be her friend; . . . So to bed, and did lie now a little better then formerly, with but little and yet with some trouble.

13 November. I home, and there to talk, with great pleasure, all the evening with my wife, who tells me that Deb hath been abroad today, and is come home and says she hath got a place to go to, so as she will be gone tomorrow morning. This troubled me; and the truth is, I have a great mind for to have the maidenhead of this girl, which I should not doubt to have if yo could get time para be con her—but she will be gone and I know not whither.

14 November. Up, and had a mighty mind to have seen or given a note to Deb or to have given her a little money; to which purpose I wrapped up 40*s* in a paper, thinking to give her; but my wife rose presently, and would not let me be out of her sight; and went down before me into the kitchen, and came up and told me that she was in the kitchen, and therefore would have me go round the other way; which she repeating, and I vexed at it, answered her a little angrily; upon which she instantly flew out into a rage, calling me dog and rogue, and that I had a rotten heart; all which, knowing that I deserved it, I bore with; and word being brought presently up that she was gone away by coach with her things, my wife was friends; and so all quiet, and I to the office with my heart sad, and find that I cannot forget the girl, and vexed I know not where to look for her. It will be I fear a little time before I shall be able to wear Deb out of my mind.

I must here remember that I have laid with my moher as a husband more times since this falling-out then in I believe twelve months before—and with more pleasure to her then I think in all the time of our marriage before.

18 November. Lay long in bed, talking with my wife, she being unwilling to have me go abroad, being and declaring herself jealous of my going out, for fear of my going to Deb; which I do deny—for which God forgive me, for I was no sooner out about noon but I did go by coach directly to Somerset House and there enquired among the porters . . .

So I could not be commanded by my reason, but I must go this very night; and so by coach, it being now dark, I to her, close by my tailor's; and there she came into the coach to me, and yo did besar her and tocar her thing, but ella was against it and laboured with much earnestness, such as I believed to be real. I did nevertheless give her the best counsel I could, to have a care of her honour and to fear God and suffer no man para haver to do con her as yo have done—which she promised. Yo did give her 20*s* and directions para laisser sealed in paper at any time the name of the place of her being, at Herringman's my bookseller in the Change—by which I might go para her. And so home, and there told my wife a fair tale, God knows, how I spent the whole day; with which the poor wretch was satisfied, or at least seemed so; and so to supper and to bed, she having been mighty busy all day in getting of her house in order against tomorrow, to hang up our new hangings and furnishing our best chamber.

19 November. Up, and at the office all the morning, with my heart full of joy to think in what a safe condition all my matters now stand between my wife and Deb and me; and at noon, running upstairs to see the upholsters, who are at work upon hanging my best room and setting up my new bed, I find my wife sitting sad in the dining room; which inquiring into the reason of, she begun to call me all the false, rotten-hearted rogues in the world, letting me understand that I was with Deb yesterday; which, thinking impossible for her ever to understand, I did a while deny; but at last did, for the ease of my mind and hers, and for ever to discharge my heart of this wicked business, I did confess all; and above stairs in our bedchamber there I did endure the sorrow of her threats and vows and curses all the afternoon. And which was worst, she swore by all that was good that she would slit the nose of this girl, and be gone herself this very night from me; and did there demand 3 or 400*l* of me to buy my peace, that she might be gone without making any noise, or else protested that she would make all the world know of it. So, with most perfect confusion of face and heart, and sorrow and shame, in the greatest agony in the world, I did pass this afternoon, fearing that it will never have an end; . . . but at last I did call for W. Hewers, who I was forced to make privy now to all; and the poor fellow did cry like a child [and] obtained what I could not, that she would be pacified, upon condition that I would give it under my hand never to see or speak with Deb while I live, . . .

So before it was late, there was, beyond my hopes as well as desert, a tolerable peace; and so to supper, and pretty kind words, and to bed, and there yo did hazer con ella to her content; and so with some rest spent the night in bed, being most absolutely resolved, if ever I can maister this bout, never to give her occasion while I live of more trouble of this or any other kind, there being no curse in the world so great as this of the difference between myself and her; and therefore I do by the grace of God promise never to offend her more, and did this night begin to pray to God upon my knees alone in my chamber; which God knows I cannot yet do heartily, but I hope God will give me the grace more and more every day to fear Him, and to be true to my poor wife. This night the upholsters did finish the hanging of my best chamber, but my sorrow and trouble is so great about this business, that put me out of all joy.

20 November. This morning up, with mighty kind words between my poor wife and I; and so to Whitehall by water, . . . But when I came home, hoping for a further degree of peace and quiet, I find my wife upon her bed in a horrible rage afresh, calling me all the bitter names; and rising, did fall to revile me in the bitterest manner in the world, and could not refrain to strike me and pull my hair; which I resolved to bear with, and had good reason to bear it. So I by silence and weeping did prevail with her a little to be quiet, and she would not eat her dinner without me; but yet by and by into a raging fit she fell again worse then before, that she would slit the girl's nose; and at last W. Hewer came in and came up, who did allay her fury, I flinging myself in a sad desperate condition upon the bed in the blue room, and there lay while they spoke together; and at last it came to this, that if I would call Deb 'whore' under my hand, and write to her that I hated her and would never see her more, she would believe me and trust in me . . .

So from that minute my wife begun to be kind to me, and we to kiss and be friends, and so continued all the evening and fell to talk of other matters with great comfort, and after supper to bed. I did this night promise to my wife never to go to bed without calling upon God upon my knees by prayer; and I begun this night, and hope I shall never forget to do the like all my life—for I do find that it is much the best for my soul and body to live pleasing to God and my poor wife—and will ease me of much care, as well as much expense.

*

29 November. Lords Day. Lay long in bed with pleasure [with my wife], with whom I have now a great deal of content; and my mind is in other things also mightily more at ease, and I do mind my business better than ever and am more at peace; and trust in God I shall ever be so, though I cannot yet get my mind off from thinking now and then of Deb.

5 December. Up, after a little talk with my wife which troubled me, she being ever since our late difference mighty watchful of sleep and dreams, and will not be persuaded but I do dream of Deb, and doth tell me that I speak in my dream and that this night I did cry 'Huzzy!' and it must be she—and now and then I start otherwise then I used to do, she says; which I know not, for I do not know that I dream of her more then usual, though I cannot deny that my thoughts waking do run now and then, against my will and judgment, upon her.

31 January. Lords Day. Lay long, talking with pleasure, and so up, and I to church. . . . And thus ended this month, with many different days of sadness and mirth, from differences between me and my wife, from her remembrance of my late unkindness to her with Willet, she not being able to forget it, but now and then hath her passionate remembrance of it, as often as prompted to it by any occasion; but this night we are at present very kind. And so ends this month.

15 April 1669. Going down Holborn Hill by the Conduit, I did see Deb on foot going up the hill; I saw her, and she me, but she made no stop, but seemed unwilling to speak to me; so I away on, but then stopped and light and after her, and overtook her at the end of Hosier Lane in Smithfield; and without standing in the street, desired her to fallow me, and I led her into a little blind alehouse within the walls; and there she and I alone fell to talk and besar la and tocar su mamelles; but she mighty coy, and I hope modest; but however, though with great force, did hazer ella con su hand para tocar mi thing, but ella was in great pain para be brought para it. I did give her in a paper 20*s*, and we did agree para meet again in the Hall at Westminster on Monday next; and so, giving me great hopes by her carriage that she continues modest and honest, we did there part . . .

SAMUEL PEPYS, *Diary*.

¶ *Pepys closed his diary in May 1669, after apparently no further contact with Deb. He probably ceased writing from concern for his failing eyesight. His wife died later that same year.*

This scene illustrates the strange attraction sometimes felt between the two rivals in a love triangle. Tereza, a photographer married to Tomas, has come to meet Sabina, his mistress:

Next to the bed stood a small table, and on the table the model of a human head, the kind hairdressers put wigs on. Sabina's wig stand sported a bowler hat rather than a wig. 'It used to belong to my grandfather,' she said with a smile.

It was the kind of hat—black, hard, round—that Tereza had seen only on the screen, the kind of hat Chaplin wore. She smiled back, picked it up, and after studying it for a time, said, 'Would you like me to take your picture in it?'

Sabina laughed for a long time at the idea. Tereza put down the bowler hat, picked up her camera, and started taking pictures.

When she had been at it for almost an hour, she suddenly said, 'What would you say to some nude shots?'

'Nude shots?' Sabina laughed.

'Yes,' said Tereza, repeating her proposal more boldly, 'nude shots.'

'That calls for a drink,' said Sabina, and opened a bottle of wine.

Tereza felt her body going weak; she was suddenly tongue-tied. Sabina, meanwhile, strode back and forth, wine in hand, going on about her grandfather, who'd been the mayor of a small town . . .

Sabina went on and on about the bowler hat and her grandfather until, emptying her third glass, she said 'I'll be right back' and disappeared into the bathroom.

She came out in her bathrobe. Tereza picked up her camera and put it to her eye. Sabina threw open the robe.

*

The camera served Tereza as both a mechanical eye through which to observe Tomas's mistress and a veil by which to conceal her face from her.

It took Sabina some time before she could bring herself to slip out of the robe entirely. The situation she found herself in was proving a bit more difficult than she had expected. After several minutes of posing, she went up to Tereza and said, 'Now it's my turn to take your picture. Strip!'

Sabina had heard the command 'Strip!' so many times from Tomas that it was engraved in her memory. Thus, Tomas's mistress had just given Tomas's command to Tomas's wife. The two women were joined by the same magic word. That was Tomas's way of unexpectedly turning an innocent conversation with a woman into an erotic situation. Instead of stroking, flattering, pleading, he would issue a command, issue it abruptly, unexpectedly, softly yet firmly and authoritatively, and at a distance: at such moments he never touched the woman he was addressing. He often used it on Tereza as well, and even though he said it softly, even though he whispered it, it was a command, and obeying never failed to arouse her. Hearing the word now made her desire to obey even stronger, because doing a stranger's bidding is a special madness, a madness all the more heady in this case because the command came not from a man but from a woman.

Sabina took the camera from her, and Tereza took off her clothes. There she stood before Sabina naked and disarmed. Literally *disarmed*: deprived of the apparatus she had been using to cover her face and aim at Sabina like a weapon. She was completely at the mercy of Tomas's mistress. This beautiful submission intoxicated Tereza. She wished that the moments she stood naked opposite Sabina would never end.

I think that Sabina, too, felt the strange enchantment of the situation: her lover's wife standing oddly compliant and timorous before her. But after clicking the shutter two or three times, almost frightened by the enchantment and eager to dispel it, she burst into loud laughter.

Tereza followed suit, and the two of them got dressed.

MILAN KUNDERA, *The Unbearable Lightness of Being*, 1984

Edith Wharton shows how an idealized third person can sometimes help to keep a marriage intact:

'We're damnably dull. We've no character, no colour, no variety.—I wonder,' he broke out, 'why you don't go back?'

Her eyes darkened, and he expected an indignant rejoinder. But she sat silent, as if thinking over what he had said, and he grew frightened lest she should answer that she wondered too.

At length she said: 'I believe it's because of you.'

It was impossible to make the confession more dispassionately, or in a tone less encouraging to the vanity of the person addressed. Archer reddened to the temples, but dared not move or speak: it was as if her words had been some rare butterfly that the least motion might drive off on startled wings, but that might gather a flock about it if it were left undisturbed.

'At least,' she continued, 'it was you who made me understand that under the dullness there are things so fine and sensitive and delicate that even those I most cared for in my other life look cheap in comparison. I don't know how to explain myself'—she drew together her troubled brows—'but it seems as if I'd never before understood with how much that is hard and shabby and base the most exquisite pleasures may be paid for.'

'Exquisite pleasures—it's something to have had them!' he felt like retorting; but the appeal in her eyes kept him silent.

'I want,' she went on, 'to be perfectly honest with you—and with myself. For a long time I've hoped this chance would come: that I might tell you how you've helped me, what you've made of me—'

Archer sat staring beneath frowning brows. He interrupted her with a laugh. 'And what do you make out that you've made of me?'

She paled a little. 'Of you?'

'Yes: for I'm of your making much more than you ever were of mine. I'm the man who married one woman because another one told him to.'

Her paleness turned to a fugitive flush. 'I thought—you promised—you were not to say such things today.'

'Ah—how like a woman! None of you will ever see a bad business through!'

She lowered her voice. '*Is* it a bad business—for May?'

He stood in the window, drumming against the raised sash, and feeling in every fibre the wistful tenderness with which she had spoken her cousin's name.

'For that's the thing we've always got to think of—haven't we—by your own showing?' she insisted.

'My own showing?' he echoed, his blank eyes still on the sea.

'Or if not,' she continued, pursuing her own thought with a painful application, 'if it's not worth while to have given up, to have missed things, so that others may be saved from disillusionment and misery—then everything I came home for, everything that made my other life seem by contrast so bare and so poor because no one there took account of them—all these things are a sham or a dream—'

He turned around without moving from his place. 'And in that case there's no reason on earth why you shouldn't go back?' he concluded for her.

Her eyes were clinging to him desperately. 'Oh, *is* there no reason?'

'Not if you staked your all on the success of my marriage. My marriage,' he said savagely, 'isn't going to be a sight to keep you here.' She made no answer, and he went on: 'What's the use? You gave me my first glimpse of a real life, and at the same moment you asked me to go on with a sham one. It's beyond human enduring—that's all.'

'Oh, don't say that; when I'm enduring it!' she burst out, her eyes filling.

Her arms had dropped along the table, and she sat with her face abandoned to his gaze as if in the recklessness of a desperate peril. The face exposed her as much as if it had been her whole person, with the soul behind it: Archer stood dumb, overwhelmed by what it suddenly told him.

'You too—oh, all this time, you too?'

For answer, she let the tears on her lids overflow and run slowly downward.

Half the width of the room was still between them, and neither made any show of moving. Archer was conscious of a curious indifference to her bodily presence: he would hardly have been aware of it if one of the hands she had flung out on the table had not drawn his gaze as on the occasion when, in the little Twenty-third Street house, he had kept his eye on it in order not to look at her face. Now his imagination spun about the hand as about the edge of a vortex; but still he made no effort to draw nearer. He had known the love that is fed on caresses and feeds them; but this passion that was closer than his bones was not to be superficially satisfied. His one terror was to do anything which might efface the sound and impression of her words; his one thought, that he should never again feel quite alone.

But after a moment the sense of waste and ruin overcame him. There they were, close together and safe and shut in; yet so chained to their separate destinies that they might as well have been half the world apart.

'What's the use—when you will go back?' he broke out, a great hopeless *How on earth can I keep you?* crying out to her beneath his words.

She sat motionless, with lowered lids. 'Oh—I shan't go yet!'

'Not yet? Some time, then? Some time that you already foresee?'

At that she raised her clearest eyes. 'I promise you: not as long as you hold out. Not as long as we can look straight at each other like this.'

He dropped into his chair. What her answer really said was: 'If you lift a finger you'll drive me back: back to all the abominations you know of, and all the temptations you half guess.' He understood it as clearly as if she had uttered the words, and the thought kept him anchored to his side of the table in a kind of moved and sacred submission.

'What a life for you!—' he groaned.

'Oh—as long as it's a part of yours.'

'And mine a part of yours?'

She nodded.

'And that's to be all—for either of us?'

'Well; it *is* all, isn't it?'

At that he sprang up, forgetting everything but the sweetness of her face. She rose too, not as if to meet him or to flee from him, but quietly, as though the worst of the task were done and she had only to wait; so quietly that, as he came closer, her outstretched hands acted not as a check but as a guide to him. They fell into his, while her arms, extended but not rigid, kept him far enough off to let her surrendered face say the rest.

They may have stood in that way for a long time, or only for a few moments; but it was long enough for her silence to communicate all she had to say, and for him to feel that only one thing mattered. He must do nothing to make this meeting their last; he must leave their future in her care, asking only that she should keep fast hold of it.

'Don't—don't be unhappy,' she said, with a break in her voice, as she drew her hands away; and he answered: 'You won't go back—you won't go back?' as if it were the one possibility he could not bear.

'I won't go back,' she said; and turning away she opened the door and led the way into the public dining-room.

*

Once more on the boat, and in the presence of others, Archer felt a tranquillity of spirit that surprised as much as it sustained him.

The day, according to any current valuation, had been a rather ridiculous failure; he had not so much as touched Madame Olenska's hand with his lips, or extracted one word from her that gave promise of further opportunities. Nevertheless, for a man sick with unsatisfied love, and parting for an indefinite period from the object of his passion, he felt himself almost humiliatingly calm and comforted. It was the perfect balance she had held between their loyalty to others and their honesty to themselves that had so stirred and yet tranquillized him; a balance not artfully calculated, as her tears and her falterings showed, but resulting naturally from her unabashed sincerity. It filled him with a tender awe, now the danger was over, and made him thank the fates that no personal vanity, no sense of playing a part before sophisticated witnesses, had tempted him to tempt her. Even after they had clasped hands for good-bye at the Fall River station, and he had turned away alone, the conviction remained with him of having saved out of their meeting much more than he had sacrificed. . . .

It was clear to him, and it grew more clear under closer scrutiny, that if she should finally decide on returning to Europe—returning to her husband—it would not be because her old life tempted her, even on the new terms offered. No: she would go only if she felt herself becoming a temptation to Archer, a temptation to fall away from the standard they had both set up. Her choice would be to stay near him as long as he did not ask her to come nearer; and it depended on himself to keep her just there, safe but secluded.

EDITH WHARTON, *The Age of Innocence*, 1920

A LOVE LETTER: RUTH ARBEITER TO MAJOR PAUL MAXWELL

September 3, 1945 Clearfield, Vermont

Dearest,

You must know that I think of you continually,
often entering unexpectedly
that brighter isolate planet where we two live.
Which resembles this earth—its air,
grass, houses, beds, laundry, things to eat—
except that it is articulate,
the accessory, understanding, speaking of
where we are born and love and
move together continually.

Departures are dreams of home,
returns to bodies and minds we're in the habit of.
And what are these terrible things
they are taking for granted? Air and grass,
houses and beds, laundry and things to eat—
so little clarity, so little space between them,
a crowd of distractions to be
bought and done and arranged for,
drugs for the surely incurable pain of
living misunderstood among many who love you.

One evening not long ago
I walked to the high flat stone where,
as children, we used to lie in wait
for the constellations. It was dusk and hazy,
the hills, soft layers of differentiated shadow
thick with the scraping of crickets, or katydids,
or whatever you call those shrill unquenchable insects,
sawing their way through night after summer night.
Seated on the stone,
straining into the distance,
it was strangely as if I were
seeing through sound. As if an intensity,

a nagging around me, somehow *became* the mist—
the hills, too, breathing quietly, the sun
quietly falling, disappearing through gauze.

Such seeings have occurred frequently
since we were together. Your quality of perceiving,
a way through, perhaps, or out of, this
damaging anguish. As when we looked down—
you remember that day—into the grassy horse pool
where one bull frog and one crimson maple leaf
quietly brought our hands together.

Dearest, what more can I say?
Here, among my chores and my children.
Mine and my husband's children. So many friends.
And in between, these incredible perspectives,
openings entirely ours in the eddying numbness
where, as you know, I am waiting for you
continually.

Ruth

ANNE STEVENSON, *Correspondences*, 1974

And he had made her smaller. The wife who is betrayed for a grand passion retains some of her dignity. Pale-faced and silent, or even storming and wailing as in classical drama, she has a tragic authority. She too has been the victim of a natural disaster, an act of the gods. But if she was set aside merely for some trivial, carnal impulse, her value also must be trivial.

What is so awful, so unfair, is that identity is at the mercy of circumstances, of other people's actions. Brian, by committing casual adultery, had turned Erica into the typical wife of a casually unfaithful husband: jealous and shrewish and unforgiving—and also, since she had been so easily deceived, dumb and insensitive.

*

For years they were moral and social allies; together they observed and judged the world. Now she judges him. They judge each other, and each finds the other guilty.

Yes, perhaps, Brian thinks, standing among the lettuces. But he had committed no overt act of aggression against Erica, deprived her of nothing. He had held to the Kennanite principle of containment, of separate spheres of action. Within the family, the marital sphere, he had been faithful. The idea of sleeping with Wendy in the marital bedroom, even if it could have been done with absolute safety, revolted him.

And even if he is guilty, he is guilty of adultery, a form of love. Erica is guilty of unforgiveness, a form of hate. Besides, his crime is over; hers continues. Three months have passed; but still in every look, every gesture, Erica shows that she has not forgotten, has not pardoned him.

It is as if he has incurred a debt which his wife will never let him repay, yet which she does not wish to forgive. She likes to see me in the wrong, Brian thinks, looking across the dark lawn at Erica; she intends to keep me there, possibly for the rest of my life.

Very well. If he is to be imprisoned for life in the wrong, why should it be a solitary confinement? Let him have some company there, the company of a warm and willing fellow-criminal. Or, to change the metaphor; if he is to be hanged for his crime, he might as well be hanged for a ram as a lamb.

*

It is night now. Brian turns a page; its shadow flaps slowly across the table. Hearing another sigh, he looks up at his wife. She is staring into the middle distance out of eyes circled in muted blue.

Now Erica turns her head. For a moment their eyes meet; then both look down. Erica knows that Brian knows what she is thinking about, and he knows she knows he knows. This mutual knowledge is like a series of infinitely disappearing, darkening ugly reflections in two opposite mirrors. But if he asks what she is thinking, she will not admit it. She knows that he does not want to ask her anyhow; he does not want to bring up the subject again. And she knows she must not bring it up.

So they say nothing. There is nothing to say.

ALISON LURIE, *The War Between the Tates*, 1974

—

On reaching home, and entering the little lighted hall with his latchkey, the first thing that caught his eye was his wife's gold-mounted umbrella lying on the rug chest. Flinging off his fur coat, he hurried to the drawing-room.

The curtains were drawn for the night, a bright fire of cedar logs burned in the grate, and by its light he saw Irene sitting in her usual corner on the sofa. He shut the door softly, and went towards her. She did not move, and did not seem to see him.

'So you've come back?' he said. 'Why are you sitting here in the dark?'

Then he caught sight of her face, so white and motionless that it seemed as though the blood must have stopped flowing in her veins; and her eyes, that looked enormous, like the great, wide, startled brown eyes of an owl. . . .

Huddled in her grey fur against the sofa cushions, she had a strange resemblance to a captive owl, bunched in its soft feathers against the wires of a cage. The supple erectness of her figure was gone, as though she had been broken by cruel exercise; as though there were no longer any reason for being beautiful, and supple, and erect.

'So you've come back,' he repeated.

She never looked up, and never spoke, the firelight playing over her motionless figure.

Suddenly she tried to rise, but he prevented her; it was then that he understood.

She had come back like an animal wounded to death, not knowing where to turn, not knowing what she was doing. The sight of her figure, huddled in the fur, was enough.

He knew then for certain that Bosinney had been her lover; knew that she had seen the report of his death—perhaps, like himself, had bought a paper at the draughty corner of a street, and read it.

She had come back then of her own accord, to the cage she had pined to be free of—and taking in all the tremendous significance of this, he longed to cry: 'Take your hated body, that I love, out of my house! Take away that pitiful white face, so cruel and soft—before I crush it. Get out of my sight; never let me see you again!'

And, at those unspoken words, he seemed to see her rise and move away, like a woman in a terrible dream, from which she was fighting to awake—rise and go out into the dark and cold, without a thought of him, without so much as the knowledge of his presence.

Then he cried, contradicting what he had not yet spoken, 'No; stay there!' And turning away from her, he sat down in his accustomed chair on the other side of the hearth.

They sat in silence.

And Soames thought: 'Why is all this? Why should I suffer so? What have I done? It is not my fault!'

Again he looked at her, huddled like a bird that is shot and dying, whose poor breast you see panting as the air is taken from it, whose poor eyes look at you who have shot it, with a slow, soft, unseeing look, taking farewell of all that is good—of the sun, and the air, and its mate.

So they sat, by the firelight, in the silence, one on each side of the hearth.

JOHN GALSWORTHY, *A Man of Property*, 1906

Lady Dedlock had a love-affair before her marriage. This has just been revealed to her husband by his lawyer, who is subsequently found murdered:

She hurriedly addresses these lines to her husband, seals, and leaves them on her table.

'If I am sought for, or accused of, his murder, believe that I am wholly innocent. Believe no other good of me, for I am innocent of nothing else that you have heard, or will hear, laid to my charge. He prepared me, on that fatal night, for his disclosure of my guilt to you. After he had left me, I went out, on pretence of walking in the garden where I sometimes walk, but really to follow him, and make one last petition that he would not protract the dreadful suspense on which I have been racked by him, you do not know how long, but would mercifully strike next morning.

I found his house dark and silent. I rang twice at his door, but there was no reply, and I came home.

I have no home left. I will encumber you no more. May you, in your just resentment, be able to forget the unworthy woman on whom you have wasted a most generous devotion—who avoids you, only with a deeper shame than that with which she hurries from herself—and who writes this last adieu.'

She veils and dresses quickly, leaves all her jewels and her money, listens, goes downstairs at a moment when the hall is empty, opens and shuts the great door; flutters away, in the shrill frosty wind.

*

'You find me,' says Sir Leicester, whose eyes are much attracted towards him, 'far from well, George Rouncewell.'

'I am very sorry both to hear it and to see it, Sir Leicester.'

'I am sure you are. No. In addition to my older malady, I have had a sudden and bad attack. Something that deadens—' making an endeavour to pass one hand down one side; 'and confuses—' touching his lips.

*

'I was about to add,' he presently goes on, 'I was about to add, respecting this attack, that it was unfortunately simultaneous with a slight misunderstanding between my Lady and myself. I do not mean that there was any difference between us (for there has been none), but that there was a misunderstanding of certain circumstances important only to ourselves, which deprives me, for a little while, of my Lady's society. She has found it necessary to make a journey—I trust will shortly return. Volumnia, do I make myself intelligible? The words are not quite under my command, in the manner of pronouncing them.' . . .

'Therefore, Volumnia, I desire to say in your presence—and in the presence of my old retainer and friend, Mrs Rouncewell, whose truth and fidelity no one can question—and in the presence of her son, George, who comes back like a familiar recollection of my youth in the home of my ancestors at Chesney Wold—in case I should relapse, in case I should not recover, in case I should lose both my speech and the power of writing, though I hope for better things—'

The old housekeeper weeping silently; Volumnia in the greatest agitation, with the freshest bloom on her cheeks; the trooper with his arms folded and his head a little bent, respectfully attentive.

'Therefore I desire to say, and to call you all to witness—beginning, Volumnia, with yourself, most solemnly—that I am on unaltered terms with Lady Dedlock. That I assert no cause whatever of complaint against her. That I have ever had the strongest affection for her, and that I retain it undiminished. Say this to herself, and to every one. If you ever say less than this, you will be guilty of deliberate falsehood to me.'

Volumnia tremblingly protests that she will observe his injunctions to the letter.

'My Lady is too high in position, too handsome, too accomplished, too superior in most respects to the best of those by whom she is surrounded, not to have her enemies and traducers, I dare say. Let it be known to them, as I make it known to you, that being of sound mind, memory, and understanding, I revoke no disposition I have made in her favour. I abridge nothing I have ever bestowed upon her. I am on unaltered terms with her, and I recall—having the full power to do it if I were so disposed, as you see—no act I have done for her advantage and happiness.'

His formal array of words might have at any other time, as it has often had, something ludicrous in it; but at this time it is serious and affecting. His noble earnestness, his fidelity, his gallant shielding of her, his generous conquest of his own wrong and his own pride for her sake, are simply honourable, manly, and true.

CHARLES DICKENS, *Bleak House*, 1852–3

8

Misery, Mayhem, and Murder

'How good of God', said Samuel Butler of Thomas and Jane Carlyle, 'to let the Carlyles marry each other, and so make only two people miserable instead of four.'

Some couples seem dedicated to misery; for others a quarrel, especially the first, is a disaster, 'as if a fracture in delicate crystal had begun, and he was afraid of any movement that might make it fatal' (George Eliot, *Middlemarch*).

Quarrelling can be a ritual: the combatants take up their familiar corners, and heaven help anyone who dares to intervene. 'Marriage resembles a pair of shears, so joined that they cannot be separated; often moving in opposite directions, yet always punishing anyone who comes between them' (Sydney Smith).

There is usually a moment of choice, when one or both partners recognize that the opening moves have been made, but the fight could still be averted.

As soon as Natasha and Pierre were alone they too began to talk as only husband and wife can talk—that is, exchanging ideas with extraordinary swiftness and perspicuity, by a method contrary to all the rules of logic, without the aid of premisses, deductions or conclusions, and in a quite peculiar way. Natasha was so used to talking to her husband in this fashion that a logical sequence of thought on Pierre's part was to her an infallible sign of something being wrong between them. When he began proving anything or calmly arguing, and she, led on by his example, began to do the same, she knew that they were on the verge of a quarrel.

(Leo Tolstoy, *War and Peace*)

Even in such idyllic marriages as Natasha and Pierre's, quarrels will—and should—occur: a marriage without any disagreement or anger may be in danger of death by inertia, or of a cataclysmic explosion.

John Osborne's hero tries his wife's patience to the limit, like a tiresome child. Only in childhood and in marriage can we hope that our worst rages will be contained, and some battles are really a testing of the relationship.

There are also couples, like Albee's, for whom fighting is a form of closeness, a substitute for sex. 'Even to show one's anger, one's hate, one's complete lack of inhibition is taken for intimacy, and this may explain the perverted attraction married couples often have for each other, who seem intimate only when they are in bed or when they give vent to their mutual hate and rage' (Erich Fromm, *The Art of Loving*).

Marriages may exist for mutual punishment, the partner chosen precisely because the relationship will be punitive. And there are power struggles that take on a deadly viciousness: psychic as well as physical murder can be committed within marriage.

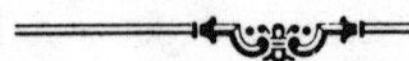

We see . . . how few matrimonies there be without chidings, brawlings, tauntings, repentings, bitter cursings, and fightings.

'On the State of Matrimony', *Book of Homilies*, 1571

Many are the Epistles I every Day receive from Husbands, who complain of Vanity, Pride, but above all Ill-nature, in their Wives. I cannot tell how it is, but I think I see in all their Letters that the Cause of their Uneasiness is in themselves; and indeed I have hardly ever observed the married Condition unhappy, but from want of Judgment or Temper in the Man. The Truth is, we generally make Love in a Stile, and with Sentiments very unfit for ordinary Life: They are half Theatrical, half Romantick. By this Means we raise our Imaginations to what is not to be expected in humane Life; and because we did not before-hand think of the Creature we were enamoured of as subject to Dishumour, Age, Sickness, Impatience, or Sullenness, but altogether considered her as the Object of Joy, humane Nature it self is often imputed to her as her particular Imperfection or Defect.

I take it to be a Rule proper to be observed in all Occurrences of Life, but more especially in the domestick or matrimonial Part of it, to preserve always a Disposition to be pleased.

*

All who are married without this Relish of their Circumstance, are either in a tasteless Indolence and Negligence, which is hardly to be attain'd, or else live in the hourly Repetition of sharp Answers, eager Upbraidings, and distracting Reproaches. In a Word, the married State, with and without the Affection suitable to it, is the compleatest Image of Heaven and Hell we are capable of receiving in this Life.

The Spectator, no. 479, 9 September 1712

The Quadrangle Club
Chicago
[March 1950]

Cat: my cat: If only you would write to me: My love, oh Cat. This is not, as it seems from the address above, a dive, joint, saloon, etc, but the honourable & dignified headquarters of the dons of the University of Chicago. I love you. That is all I know. But all I know, too, is that I am writing into space: the kind of dreadful, unknown space I am just going to enter. I am going to Iowa, Illinois, Idaho, Indindiana, but these, though mis-spelt, *are* on the map. You are not. Have you forgotten me? I am the man you used to say you loved. I used to sleep in your arms—do you remember? But you never write. You are perhaps mindless of me. I am not of you. I love you. There isn't a moment of any hideous day when I do not say to myself, 'It will be alright. I shall go home. Caitlin loves me. I love Caitlin.' But perhaps you have forgotten. If you have forgotten, or lost your affection for me, please, my Cat, let me know. I Love You.

Dylan.

DYLAN THOMAS, letter to his wife Caitlin.

Anton Chekhov married the actress Olga Leonardovna Knipper on 25 May 1901. He was already very ill with consumption, and on his doctor's orders compelled to spend the winters in Yalta, while she continued with her career.

Aug. 27 (1902. Yalta).

My darling, my perch, after long suspense, at last I have received a letter from you. I am leading a quiet life. I don't go into the town, I talk to visitors and from time to time write a little. I am not going to write a play this year. I have no heart for it. . . .

There is a chilly feeling about your letters, and yet I go on pestering you with endearments, and think of you endlessly. I kiss you a billion times and hug you. Write to me, darling, oftener than once in five days. You know I am your husband anyway. Don't part from me so soon without having lived with me properly, without having borne me a little boy or girl, and when once you have a baby, then you can behave exactly as you like. I kiss you again.

Your A——.

Sept. 1 (1902. Yalta).

My dear, my own, again I have had a strange letter from you. Again you hurl all sorts of things at my head. . . . You charge me with not being open and meanwhile you forget everything I tell you or write you. And I simply can't imagine what I am to do with my wife, how to write to her. You write that you shudder when you read my letters, that it is time for us to part, that there is something you don't understand in it all. . . . You write that I am capable of living beside you and always being silent, that I only want you as an agreeable woman and that you yourself as a human being are living lonely and a stranger to me. My sweet, good darling, but you are my wife, you know, do understand that at last. You are the person nearest and dearest to me. I have loved you infinitely and I love you still, and you go writing about being an agreeable woman, lonely and a stranger to me.

Olga wrote:

Oct. 6 (1903)

My dear, I simply don't know what to write to you. I know that you are ill, that I am for you simply a nonentity, who comes, stays with you and goes away. There is such a horrible falsity in my life that I don't know how to live . . . and here I go about desolate, I scourge myself, I blame myself on every side, I feel I am in fault all round. There is something I cannot cope with in life . . . of course you must not come to Moscow at all.

He replied:

Oct. 9 (1903. Yalta).

My pony, don't write me angrily dismal letters, don't forbid me to come to Moscow. In any case I shall come to Moscow, and if you don't let me come to you I shall stay somewhere in an hotel. You see I want very little in Moscow—in the way of conveniences—a seat at the theatre and a spacious *vaterkloset*.

Nov. 29 (1903. Yalta).

I really don't know what to do and what to think, pony. You persistently don't ask me to come to Moscow and apparently don't intend to ask me. You should write openly and tell me why it is, what is the reason, and I would not waste time, I would go abroad. If only you knew how depressingly the rain patters on the roof, how I long for a sight of my wife. But have I a wife? Where is she?

March 12 (1904. Yalta).

. . . I write this not knowing where you are or how you are: what am I to think of your silence, to what address am I to write—to Leontyevsky or to Petrovka?—I am beginning to wonder whether I hadn't better dash off to Moscow. Why, oh why, haven't you sent me one telegram about your health? Why? Evidently I count for nothing to you, I am simply superfluous. In fact, it is beastly.

from *The Letters of Anton Pavlovitch Tchehov to Olga Leonardovna Knipper*, 1924

¶ *Chekhov died on 2 July 1904 in Badenweiler, Germany, with his wife by his side. She later wrote: 'The impression that those six years [the years she had known him] have left is one of anxiety, of rushing from place to place—like a seagull over the ocean, not knowing where to alight.'*

Two people continue to see each other for a long time from habit, and to repeat that they love each other when their manner shows this is no longer true.

The beginning and the decline of love are evidenced by the embarrassment felt by the couple when they find themselves alone.

LA BRUYÈRE, *Characters*, 1688

TALKING IN BED

Talking in bed ought to be easiest,
Lying together there goes back so far,
An emblem of two people being honest.

Yet more and more time passes silently.
Outside, the wind's incomplete unrest
Builds and disperses clouds about the sky,

And dark towns heap up on the horizon.
None of this cares for us. Nothing shows why
At this unique distance from isolation

It becomes still more difficult to find
Words at once true and kind,
Or not untrue and not unkind.

PHILIP LARKIN, 1964

'You are old,' said the youth, 'and your jaws are too weak
For anything tougher than suet;
Yet you finished the goose, with the bones and the beak—
Pray, how did you manage to do it?'

'In my youth,' said his father, 'I took to the law,
And argued each case with my wife;
And the muscular strength, which it gave to my jaw,
Has lasted the rest of my life.'

LEWIS CARROLL, *Alice's Adventures in Wonderland*, 1865

DORINDA: Morrow, my dear Sister; are you for Church this Morning?

MRS SULLEN: Any where to Pray; for Heaven alone can help me: but, I think, *Dorinda*, there's no Form of Prayer in the Liturgy against bad Husbands.

DORINDA: But there's a Form of Law in *Doctors-Commons*; and I swear, Sister *Sullen*, rather than see you thus continually discontented, I would advise you to apply to that: For besides the part that I bear in your vexatious Broils, as being Sister to the Husband, and Friend to the Wife; your Example gives me such an Impression of Matrimony, that I shall be apt to condemn my Person to a long Vacation all its Life.—But supposing, Madam, that you brought it to a Case of Separation, what can you urge against your Husband? My Brother is, first, the most constant Man alive.

MRS SULLEN: The most constant Husband, I grant'ye.

DORINDA: He never sleeps from you.

MRS SULLEN: No, he always sleeps with me.

DORINDA: He allows you a Maintenance suitable to your Quality.

MRS SULLEN: A Maintenance! do you take me, Madam, for an hospital Child, that I must sit down, and bless my Benefactors for Meat, Drink and Clothes? As I take it, Madam, I brought your Brother Ten thousand Pounds, out of which, I might expect some pretty things, call'd Pleasures.

DORINDA: You share in all the Pleasures that the Country affords.

MRS SULLEN: Country Pleasures! Racks and Torments! dost think, Child, that my Limbs were made for leaping of Ditches, and clambring over Stiles; or that my Parents wisely foreseeing my future Happiness in Country-pleasures, had early instructed me in the rural Accomplishments of drinking fat Ale, playing at Whisk, and smoaking Tobacco with my Husband; or of spreading of Plaisters, brewing of Diet-drinks, and stilling Rosemary-Water with the good old Gentlewoman, my Mother-in-Law?

DORINDA: I'm sorry, Madam, that it is not more in our power to divert you; I cou'd wish indeed that our Entertainments were a little more polite, or your Taste a little less refin'd: But, pray, Madam, how came the Poets and Philosophers that labour'd so much in hunting after Pleasure, to place it at last in a Country Life?

MRS SULLEN: Because they wanted Money, Child, to find out the Pleasures of the Town . . . But yonder I see my *Coridon*, and a sweet Swain it is, Heaven knows.—Come, *Dorinda*, don't be angry, he's my Husband, and your Brother; and between both is he not a sad Brute?

DORINDA: I have nothing to say to your part of him, you're the best Judge.

MRS SULLEN: O Sister, Sister! if ever you marry, beware of a sullen, silent Sot, one that's always musing, but never thinks:—There's some Diversion in a talking Blockhead; and since a Woman must wear Chains, I wou'd have the Pleasure of hearing 'em rattle a little.—Now you shall see, but take this by the way;—He came home this Morning at his usual Hour of Four, waken'd me out of a sweet Dream of something else, by tumbling over the Tea-table, which he broke all to pieces, after his Man and he had rowl'd about the Room like sick Passengers in a Storm . . . he comes flounce into Bed, dead as a Salmon into a Fishmonger's Basket; his Feet cold as Ice, his Breath hot as a Furnace, and his Hands and his Face as greasy as his Flanel Night-cap.—Oh Matrimony!—He tosses up the Clothes with a barbarous swing over his Shoulders, disorders the whole Oeconomy of my Bed, leaves me half naked, and my whole Night's Comfort is the tuneable Serenade of that wakeful Nightingale, his Nose.—O the Pleasure of counting the melancholly Clock by a snoring Husband!

GEORGE FARQUHAR, *The Beaux' Stratagem*, 1707

LADY TEAZLE: If you please, I'm sure I don't care how soon we leave off quarrelling, provided you'll own you were tired first.

SIR PETER: Well—then let our future contest be, who shall be most obliging.

LADY TEAZLE: I assure you, Sir Peter, good nature becomes you. You look now as you did before we were married, when you used to walk with me under the elms, and tell me stories of what a gallant you were in your youth, and chuck me under the chin, you would; and ask me if I thought I could love an old fellow, who would deny me nothing—didn't you?

SIR PETER: Yes, yes, and you were as kind and attentive—

LADY TEAZLE: Ay, so I was, and would always take your part, when my acquaintance used to abuse you, and turn you into ridicule.

SIR PETER: Indeed!

LADY TEAZLE: Ay, and when my cousin Sophy has called you a stiff, peevish old bachelor, and laughed at me for thinking of marrying one who might be my father, I have always defended you, and said, I didn't think you so ugly by any means, and that you'd make a very good sort of a husband.

SIR PETER: And you prophesied right; and we shall now be the happiest couple—

LADY TEAZLE: And never differ again?

SIR PETER: No, never—though at the same time, indeed, my dear Lady Teazle, you must watch your temper very seriously; for in all our little quarrels, my dear, if you recollect, my love, you always began first.

LADY TEAZLE: I beg your pardon, my dear Sir Peter: indeed, you always gave the provocation.

SIR PETER: Now, see, my angel! take care—contradicting isn't the way to keep friends.

LADY TEAZLE: Then, don't you begin it, my love!

SIR PETER: There, now! you—you are going on. You don't perceive, my life, that you are just doing the very thing which you know always makes me angry.

LADY TEAZLE: Nay, you know if you will be angry without any reason, my dear—

SIR PETER: There! now you want to quarrel again.

LADY TEAZLE: No, I'm sure I don't: but, if you will be so peevish—

SIR PETER: There now! who begins first?

LADY TEAZLE: Why, you, to be sure. I said nothing—but there's no bearing your temper.

SIR PETER: No, no, madam: the fault's in your own temper.

LADY TEAZLE: Ay, you are just what my cousin Sophy said you would be.

SIR PETER: Your cousin Sophy is a forward, impertinent gipsy.

LADY TEAZLE: You are a great bear, I am sure, to abuse my relations.

SIR PETER: Now may all the plagues of marriage be doubled on me, if ever I try to be friends with you any more!

LADY TEAZLE: So much the better.

SIR PETER: No, no, madam: 'tis evident you never cared a pin for me, and I was a madman to marry you—a pert, rural coquette, that had refused half the honest 'squires in the neighbourhood!

LADY TEAZLE: And I am sure I was a fool to marry you—an old dangling bachelor, who was single at fifty, only because he never could meet with any one who would have him.

SIR PETER: Ay, ay, madam; but you were pleased enough to listen to me: you never had such an offer before.

LADY TEAZLE: No! didn't I refuse Sir Tivy Terrier, who everybody said would have been a better match? for his estate is just as good as yours, and he has broke his neck since we have been married.

SIR PETER: I have done with you, madam! You are an unfeeling, ungrateful—but there's an end of everything. I believe you capable of everything that is bad. Yes, madam, I now believe the reports relative to you and Charles, madam. Yes, madam, you and Charles are, not without grounds—

LADY TEAZLE: Take care, Sir Peter! you had better not insinuate any such thing! I'll not be suspected without cause, I promise you.

SIR PETER: Very well, madam! very well! a separate maintenance as soon as you please. Yes, madam, or a divorce! I'll make an example of myself for the benefit of all old bachelors. Let us separate, madam.

LADY TEAZLE: Agreed! agreed! And now, my dear Sir Peter, we are of a mind once more, we may be the happiest couple, and never differ again, you know: ha! ha! ha! Well, you are going to be in a passion, I see, and I shall only interrupt you—so, bye! bye! [*Exit.*

RICHARD BRINSLEY SHERIDAN, *The School for Scandal*, 1777

Something unfamiliar in the aspect of the breakfast-room as glimpsed through the open door from the hall, drew him within. Hilda had at last begun to make it into 'her' room. She had brought an old writing-desk from upstairs and put it between the fireplace and the window. Edwin thought: 'Doesn't she even know the light ought to fall over the left shoulder, not over the right?' . . . He thought: 'If she'd asked me, I could have arranged it for her much better than that.' Nevertheless the idea of her being absolute monarch of the little room, and expressing her individuality in it and by it, both pleased and touched him. Nor did he at all resent the fact that she had executed her plan in secret. She must have been anxious to get the room finished for the musical evening.

Thence he passed into the drawing-room,—and was thunderstruck. The arrangement of the furniture was utterly changed, and the resemblance to a boarding-house parlour after all achieved. The piano had crossed the room; the chairs were massed together in the most ridiculous way; the sofa was so placed as to be almost useless. His anger was furious but cold. The woman had considerable taste in certain directions, but she simply did not understand the art of fixing up a room, whereas he did. . . . The woman had clearly failed to appreciate the sacredness of the *status quo*. He appreciated it himself, and never altered anything without consulting her and definitely announcing his intention to alter. She probably didn't care a fig for the *status quo*. Her conduct was inexcusable. It was an attack on vital principles. It was an outrage. Doubtless, in her scorn for the *status quo*, she imagined that he would accept the *fait accompli*. She was mistaken. With astounding energy he set to work to restore the *status quo ante*. The vigour with which he dragged and peched an innocent elephantine piano was marvellous. In less than five minutes not a trace remained of the *fait accompli*. He thought: 'This is a queer start for a musical evening!' But he was triumphant, resolute, and remorseless. He would show her a thing or two. In particular he would show that fair play had to be practised in his house.

*

He revisited the drawing-room to survey his labours. She was there. Whence she had sprung he knew not. But she was there. He caught sight of her standing by the window before entering the room.

When he got into the room he saw that her emotional excitement far surpassed his own. Her lips and her hands were twitching; her nostrils dilated and contracted; tears were in her eyes.

'Edwin,' she exclaimed very passionately, in a thick voice, quite unlike her usual clear tones, as she surveyed the furniture, 'this is really too much!'

Evidently she thought of nothing but her resentment. No consideration other than her outraged dignity would have affected her demeanour. If a whole regiment of their friends had been watching at the door, her demeanour would not have altered. The bedrock of her nature had been reached.

'It's war, this is!' thought Edwin.

He was afraid; he was even intimidated by her anger; but he did not lose his courage. The determination to fight for himself, and to see the thing through no matter what happened, was not a bit weakened. An inwardly feverish but outwardly calm vindictive desperation possessed him. He and she would soon know who was the stronger.

At the same time he said to himself:

'I was hasty. I ought not to have acted in such a hurry. Before doing anything I ought to have told her quietly that I intended to have the last word as regards furniture in this house. I was within my rights in acting at once, but it wasn't very clever of me, clumsy fool!'

Aloud he said, with a kind of self-conscious snigger:

'What's too much?'

Hilda went on:

'You simply make me look a fool in my own house, before my own son and the servants.'

'You've brought it on yourself,' said he fiercely. 'If you will do these idiotic things you must take the consequences. I told you I didn't want the furniture moved, and immediately my back's turned you go and move it. I won't have it, and so I tell you straight.' . . .

'You're a brute,' she continued, not heeding him, obsessed by her own wound. 'You're a brute!' She said it with terrifying conviction. 'Everybody knows it. Didn't Maggie warn me? You're a brute and a bully . . .

'I think you ought to apologize to me,' she blubbered. 'Yes, I really do.'

'Why should I apologize to you? You moved the furniture against my wish. I moved it against yours. That's all. You began. I didn't begin. You want everything your own way. Well, you won't have it.'

She blubbered once more:

'You ought to apologize to me.'

And then she wept hysterically.

He meditated sourly, harshly. He had conquered. The furniture was as he wished, and it would remain so. The enemy was in tears, shamed, humiliated. He had a desire to restore her dignity, partly because she was his wife and partly because he hated to see any human being beaten. . . . He did not mind apologizing to her, if an apology would give her satisfaction. He was her superior in moral force, and naught else mattered.

'I don't think I ought to apologize,' he said, with a slight laugh. 'But if you think so I don't mind apologizing. I apologize. There!' He dropped into an easy-chair.

To him it was as if he had said:

'You see what a magnanimous chap I am.'

She tried to conceal her feelings, but she was pleased, flattered, astonished. Her self-respect returned to her rapidly.

'Thank you,' she murmured, and added: 'It was the least you could do.'

At her last words he thought:

'Women are incapable of being magnanimous.'

She moved towards the door.

'Hilda,' he said.

She stopped.

'Come here,' he commanded, with gentle bluffness.

She wavered towards him.

'Come here, I tell you,' he said again.

He drew her down to him, all fluttering and sobbing and wet, and kissed her, kissed her several times; and then, sitting on his knees, she kissed him. But, though she mysteriously signified forgiveness, she could not smile; she was still far too agitated and out of control to be able to smile.

The scene was over. . . . Her broken body and soul huddled against him were agreeably wistful to his triumphant manliness. But he had had a terrible fright. And even now there was a certain mere bravado in his attitude. In his heart he was thinking:

'By Jove! has it come to this?'

The responsibilities of the future seemed too complicated, wearisome, and overwhelming. The earthly career of a bachelor seemed almost heavenly in its wondrous freedom. . . . The unexampled creature, so recently the source of ineffable romance, still sat on his knees, weighing them down. Suddenly he noticed that his head ached very badly—worse than it had ached all day.

ARNOLD BENNETT, *These Twain*, 1916

LAVINIA: Yes, a very important discovery,
Finding that you've spent five years of your life
With a man who has no sense of humour;
And that the effect upon me was
That I lost all sense of humour myself.
That's what came of always giving in to you.
EDWARD: I was unaware that you'd always given in to me.
It struck me very differently. As we're on the subject
I thought that it was I who had given in to *you*.
LAVINIA: I know what you mean by giving in to *me*:
You mean, leaving all the practical decisions
That you should have made yourself. I remember—
Oh, I ought to have realised what was coming—
When we were planning our honeymoon,
I couldn't make you say where you wanted to go . . .
EDWARD: But I wanted *you* to make that decision.
LAVINIA: But how could I tell where I wanted to go
Unless you suggested some other place first?

. . . . . .

And you were so considerate, people said;
And you thought you were unselfish. It was only passivity
You only wanted to be bolstered, encouraged. . . .
EDWARD: Encouraged? To what?
LAVINIA: To think well of yourself.
You know it was I who made you work at the Bar . . .
EDWARD: You nagged me because I didn't get enough work
And said that I ought to meet more people:
But when the briefs began to come in—
And they didn't come through any of *your* friends—
You suddenly found it inconvenient
That I should be always too busy or too tired
To be of use to you socially . . .
LAVINIA: I *never* complained.

EDWARD: No; and it was perfectly infuriating,
The way you *didn't* complain . . .

. . . . . .

LAVINIA: Well, but I tried to do something about it.
That was why I took so much trouble
To have those Thursdays, to give you the chance
Of talking to intellectual people . . .
EDWARD: You would have given me about as much opportunity
If you had hired me as your butler:
Some of your guests may have thought I *was* the butler.

. . . . . .

LAVINIA: Everything I tried only made matters worse,
And the moment you were offered something that you wanted
You wanted something else. I shall treat you very differently
In future.
EDWARD: Thank you for the warning.

. . . . . .

You say you were trying to 'encourage' me:
Then why did you always make me feel insignificant?
I may not have known what life I wanted,
But it wasn't the life you chose for me.
You wanted your husband to be *successful*,
You wanted me to supply a public background
For your kind of public life. You wished to be a hostess
For whom my career would be a support.
Well, I tried to be accommodating. But, in future,
I shall behave, I assure you, very differently.
LAVINIA: Bravo! Edward. This is surprising.
Now who could have taught you to answer back like that?
EDWARD: I have had quite enough humiliation
Lately, to bring me to the point
At which humiliation ceases to humiliate.
You get to the point at which you cease to feel
And then you speak your mind.
LAVINIA: That will be a novelty
To find that you have a mind to speak.
Anyway, I'm prepared to take you as you are.
EDWARD: You mean, you are prepared to take me
As I was, or as you think I am.
But what do you think I am?

LAVINIA: Oh, what you always were.
As for me, I'm rather a different person
Whom you must get to know.
EDWARD: This is very interesting:
But you seem to assume that you've done all the changing—
Though I haven't yet found it a change for the better.
But doesn't it occur to you that possibly
I may have changed too?

. . . . . .

The change that comes
From seeing oneself through the eyes of other people.
LAVINIA: That must have been very shattering for you.
But never mind, you'll soon get over it
And find yourself another little part to play,
With another face, to take people in.
EDWARD: One of the most infuriating things about you
Has always been your perfect assurance
That you understood me better than I understood myself.
LAVINIA: And the most infuriating thing about you
Has always been your placid assumption
That I wasn't worth the trouble of understanding.
EDWARD: So here we are again. Back in the trap,
With only one difference, perhaps—we can fight each other,
Instead of each taking his corner of the cage.
Well, it's a better way of passing the evening
Than listening to the gramophone.

T. S. ELIOT, *The Cocktail Party*, 1950

People who seem to enjoy their ill-temper have a way of keeping it in fine condition by inflicting privations on themselves. That was Mrs Glegg's way: she made her tea weaker than usual this morning and declined butter. It was a hard case that a vigorous mood for quarrelling, so highly capable of using any opportunity, should not meet with a single remark from Mr Glegg on which to exercise itself. But by and by it appeared that his silence would answer the purpose, for he heard himself apostrophised at last in that tone peculiar to the wife of one's bosom.

'Well, Mr Glegg! it's a poor return I get for making you the wife I've made you all these years. If this is the way I'm to be treated, I'd better ha' known it before my poor father died, and then, when I'd wanted a home, I should ha' gone elsewhere—as the choice was offered me.'

Mr Glegg paused from his porridge and looked up—not with any new amazement but simply with that quiet, habitual wonder with which we regard constant mysteries.

'Why, Mrs G., what have I done now?'

'Done now, Mr Glegg? *done now?* . . . I'm sorry for you.'

Not seeing his way to any pertinent answer, Mr Glegg reverted to his porridge.

'There's husbands in the world,' continued Mrs Glegg after a pause, 'as 'ud have known how to do something different to siding with everybody else against their own wives. Perhaps I'm wrong, and you can teach me better—but I've allays heard as it's the husband's place to stand by the wife, instead o' rejoicing and *triumphing* when folks insult her.'

'Now, what call have you to say that?' said Mr Glegg, rather warmly, for though a kind man, he was not as meek as Moses. 'When did I rejoice or triumph over you?'

'There's ways o' doing things worse than speaking out plain, Mr Glegg. I'd sooner you'd tell me to my face as you make light of me, than try to make out as everybody's in the right but me, and come to your breakfast in the morning, as I've hardly slept an hour this night, and sulk at me as if I was the dirt under your feet.'

'Sulk at you?' said Mr Glegg, in a tone of angry facetiousness. 'You're like a tipsy man as thinks everybody's had too much but himself.'

'Don't lower yourself with using coarse language to *me*, Mr Glegg! It makes you look very small, though you can't see yourself,' said Mrs Glegg in a tone of energetic compassion. 'A man in your place should set an example, and talk more sensible.'

'Yes; but will you listen to sense?' retorted Mr Glegg, sharply. 'The best sense I can talk to you is what I said last night—as you're i' the wrong to think o' calling in your money, when it's safe enough if you'd let it alone, all because of a bit of a tiff, and I was in hopes you'd ha' altered your mind this morning. But if you'd like to call it in, don't do it in a hurry now, and breed more enmity in the family—but wait till there's a pretty mortgage to be had without any trouble. You'd have to set the lawyer to work now to find an investment, and make no end o' expense.'

Mrs Glegg felt there was really something in this, but she tossed her head and emitted a guttural interjection to indicate that her silence was only an armistice, not a peace. And, in fact, hostilities soon broke out again.

'I'll thank you for my cup o' tea now, Mrs G.,' said Mr Glegg, seeing that she did not proceed to give it him as usual, when he had finished his porridge. She lifted the teapot with a slight toss of the head, and said,

'I'm glad to hear you'll *thank* me, Mr Glegg. It's little thanks *I* get for what I do for folks i' this world. Though there's never a woman o' *your* side i' the family, Mr Glegg, as is fit to stand up with me, and I'd say it if I was on my dying bed. Not but what I've allays conducted myself civil to your kin, and there isn't one of 'em can say the contrary, though my equils they aren't, and nobody shall make me say it.'

'You'd better leave finding fault wi' my kin till you've left off quarrelling with your own, Mrs G.,' said Mr Glegg, with angry sarcasm. 'I'll trouble you for the milk-jug.'

'That's as false a word as ever you spoke, Mr Glegg,' said the lady, pouring out the milk with unusual profuseness, as much as to say, if he wanted milk, he should have it with a vengeance. 'And you know it's false. I'm not the woman to quarrel with my own kin: *you* may, for I've known you do it.'

'Why, what did you call it yesterday, then, leaving your sister's house in a tantrum?'

'I'd no quarrel wi' my sister, Mr Glegg, and it's false to say it. Mr Tulliver's none o' my blood, and it was him quarrelled with me, and drove me out o' the house. But perhaps you'd have had me stay and be swore at, Mr Glegg; perhaps you was vexed not to hear more abuse and foul language poured out upo' your own wife. But let me tell you, it's *your* disgrace.'

'Did ever anybody hear the like i' this parish?' said Mr Glegg, getting hot. 'A woman with everything provided for her, and allowed to keep her own money the same as if it was settled on her, and with a gig new-stuffed and lined at no end o' expense, and provided for when I die beyond anything she could expect . . . to go on i' this way, biting and snapping like a mad dog! It's beyond everything as God A'mighty should ha' made women so.' (These last words were uttered in a tone of sorrowful agitation: Mr Glegg pushed his tea from him, and tapped the table with both his hands.)

'Well, Mr Glegg! if those are your feelings, it's best they should be known,' said Mrs Glegg, taking off her napkin, and folding it in an excited manner. 'But if you talk o' my being provided for beyond what I could expect, I beg leave to tell you as I'd a right to expect a many things as I don't find. And as to my being like a mad dog, it's well if you're not cried shame on by the county for your treatment of me, for it's what I can't bear, and I won't bear' . . .

Here Mrs Glegg's voice intimated that she was going to cry, and breaking off from speech, she rang the bell violently.

'Sally,' she said, rising from her chair, and speaking in rather a choked voice, 'light a fire upstairs, and put the blinds down. Mr Glegg, you'll please to order what you'd like for dinner. I shall have gruel.'

Mrs Glegg walked across the room to the small bookcase, and took down Baxter's 'Saints' Everlasting Rest' which she carried with her upstairs. It was the book she was accustomed to lay open before her on special occasions: on wet Sunday mornings—or when she heard of a death in the family—or when, as in this case, her quarrel with Mr Glegg had been set an octave higher than usual.

GEORGE ELIOT, *The Mill on the Floss*, 1860

Jimmy, determined to vent his frustration and fury with society on his wife, starts by using his friend Cliff as lightning-conductor:

JIMMY: Do you have to make all that racket?

CLIFF: Oh, sorry.

JIMMY: It's quite a simple thing, you know—turning over a page. Anyway, that's my paper. (*Snatches it away.*)

CLIFF: Oh, don't be so mean!

JIMMY: Price ninepence, obtainable from any newsagent's. Now let me hear the music, for God's sake.

Pause.

(*to Alison*). Are you going to be much longer doing that?

ALISON: Why?

JIMMY: Perhaps you haven't noticed it, but it's interfering with the radio.

ALISON: I'm sorry. I shan't be much longer.

A pause. The iron mingles with the music. Cliff shifts restlessly in his chair, Jimmy watches Alison, his foot beginning to twitch dangerously. Presently, he gets up quickly, crossing below Alison to the radio, and turns it off.

What did you do that for?

JIMMY: I wanted to listen to the concert, that's all.

ALISON: Well, what's stopping you?

JIMMY: Everyone's making such a din—that's what's stopping me.

ALISON: Well, I'm very sorry, but I can't just stop everything because you want to listen to music.

JIMMY: Why not?

ALISON: Really, Jimmy, you're like a child.

JIMMY: Don't try and patronise me. (*Turning to Cliff.*) She's so clumsy. I watch for her to do the same things every night. The way she jumps on the bed, as if she were stamping on someone's face, and draws the curtains back with a great clatter, in that casually destructive way of hers. It's like someone launching a battleship. Have you ever noticed how noisy women are? Have you? The way they kick the floor about, simply walking over it? Or have you watched them sitting at their dressing tables, dropping their weapons and banging down their bits of boxes and brushes and lipsticks?

He faces her dressing table.

I've watched her doing it night after night. When you see a woman in front of her bedroom mirror, you realise what a refined sort of a butcher she is. Did you ever see some dirty old Arab, sticking his fingers into some mess of lamb fat and gristle? Well, she's just like that. Thank God they don't have many women surgeons! Those primitive hands would have your guts out in no time. Flip! Out it comes, like the powder out of its box. Flop! Back it goes, like the powder puff on the table.

CLIFF: (*grimacing cheerfully*). Ugh! Stop it!

JIMMY: She'd drop your guts like hair clips and fluff all over the floor. You've got to be fundamentally insensitive to be as noisy and as clumsy as that. . . .

Church bells start ringing outside.

Oh, hell! Now the bloody bells have started!

He rushes to the window.

Wrap it up, will you? Stop ringing those bells! There's somebody going crazy in here! I don't want to hear them!

ALISON: Stop shouting! (*Recovering immediately.*) You'll have Miss Drury up here.

JIMMY: I don't give a damn about Miss Drury—that mild old gentlewoman doesn't fool me, even if she takes in you two. She's an old robber. She gets more than enough out of us for this place every week. Anyway, she's probably in church, (*points to the window*) swinging on those bloody bells!

Cliff goes to the window, and closes it.

CLIFF: Come on now, be a good boy. I'll take us all out, and we'll have a drink.

JIMMY: They're not open yet. It's Sunday. Remember? Anyway, it's raining.

CLIFF: Well, shall we dance?

He pushes Jimmy round the floor, who is past the mood for this kind of fooling.

Do you come here often?

JIMMY: Only in the mating season. All right, all right, very funny.

He tries to escape, but Cliff holds him like a vice.

Let me go.

CLIFF: Not until you've apologised for being nasty to everyone. Do you think bosoms will be in or out, this year?

JIMMY: Your teeth will be out in a minute, if you don't let go!

He makes a great effort to wrench himself free, but Cliff hangs on. They collapse to the floor, below the table, struggling. Alison carries on with her ironing. This is routine, but she is getting close to breaking point, all the same. Cliff manages to break away, and finds himself in front of the ironing board. Jimmy springs up. They grapple.

ALISON: Look out, for heaven's sake! Oh, it's more like a zoo every day!

Jimmy makes a frantic, deliberate effort, and manages to push Cliff on to the ironing board, and into Alison. The board collapses. Cliff falls against her, and they end up in a heap on the floor. Alison cries out in pain. Jimmy looks down at them, dazed and breathless.

CLIFF: (*picking himself up*). She's hurt. Are you all right?

ALISON: Well, does it look like it!

CLIFF: She's burnt her arm on the iron.

JIMMY: Darling, I'm sorry.

ALISON: Get out!

JIMMY: I'm sorry, believe me. You think I did it on pur—

ALISON: (*her head shaking helplessly*). Clear out of my *sight*!

He stares at her uncertainly. Cliff nods to him, and he turns and goes out of the door.

JOHN OSBORNE, *Look Back in Anger*, 1956

But in old age Queenie had him well in hand. He knew that he had to produce at least a few shillings on Saturday night, or, when Sunday dinner-time came, Queenie would spread the bare cloth on the table and they would just have to sit down and look at each other: there would be no food.

Forty-five years before she had served him with a dish even less to his taste. He had got drunk and beaten her cruelly with the strap with which he used to keep up his trousers. Poor Queenie had gone to bed sobbing; but she was not too overcome to think, and she decided to try an old country cure for such offences.

The next morning when he came to dress, his strap was missing. Probably already ashamed of himself, he said nothing, but hitched up his trousers with string and slunk off to work, leaving Queenie apparently still asleep.

At night, when he came home to tea, a handsome pie was placed before him, baked a beautiful golden-brown and with a pastry tulip on the top: such a pie as must have seemed to him to illustrate the old saying: '*A woman, a dog and a walnut tree, the more you beat 'em the better they be*.'

'You cut it, Tom,' said a smiling Queenie. 'I made it a-purpose for you. Come, dont 'ee be afraid on it. 'Tis all for you.' And she turned her back and pretended to be hunting for something in the cupboard.

Tom cut it: then recoiled, for, curled up inside, was the leather strap with which he had beaten his wife. 'A just went as white as a ghoo-ost, an' got up an' went out,' said Queenie all those years later. 'But it cured 'en, for's not so much as laid a finger on me from that day to this!'

FLORA THOMPSON, *Lark Rise to Candleford*, 1939

He had what he would call in the charges made in other people's divorce petitions, a 'violent and ungovernable temper', although I can't remember it getting any worse when he lost his sight. He would shout on railway platforms, in restaurants, in the corridors of the Law Courts where he once yelled Macbeth's curse, 'The devil damn thee black, thou cream-faced loon!' at an instructing solicitor who had forgotten a document. Cold plates, soft-boiled eggs, being kept waiting for anything: such irritations rather than the disaster of blindness would make him thunder at my mother, 'Kath! Kath! Are you a complete cretin?' Only sometimes, after long periods of abuse borne patiently, she would walk away from him and my father would be left standing in the middle of the bedroom, a silver-backed hairbrush in his hand and his braces dangling, panic-stricken and yelling, 'Kath! Kath!' into the unresponsive darkness around him.

JOHN MORTIMER, *Clinging to the Wreckage*, 1982

Ruth's brave smile faltered over the soup. Her parents-in-law stared up at her in calm and pleasant anticipation. And Ruth gazed at the three dog hairs in that greyish foam which is good mushroom soup, thickened by cream and put through the blender. . . .

'Don't let the soup get cold, Ruth,' said Bobbo, as if this was her usual habit.

'Hairs!' was all Ruth said.

'It's a nice clean dog,' said Brenda. 'We don't mind, do we, Angus?'

'Of course not,' said Angus, who did. As a child Bobbo had always wanted a dog, and Angus had always prevented him from having one.

'Can't you even keep the dog out of the soup?' asked Bobbo. It was the wrong thing to say, and he knew it as soon as it was said. He did try not to say 'can't you even' to Ruth, but it did slip out whenever he was feeling at odds with her, which of late had been more and more.

Tears appeared in Ruth's eyes. She picked up the soup tureen.

'I'll sieve it,' she said.

'What a good idea!' said Brenda. 'Then no harm's done.'

'Bring the soup back at once,' cried Bobbo. 'Don't be so silly, Ruth. It isn't a disaster. It's three dog hairs. Just pick them out.'

'But they might be the guinea pig's,' said Ruth. 'He was running along the dresser shelf.' She liked the guinea pig least of all the children's pets. Its shoulders were too hunched and its eyes too deep. It reminded her of herself.

'You're tired,' said Bobbo. 'You must be tired, or you wouldn't talk such nonsense. Sit down.'

'Let her sieve the soup, dear,' said Brenda, 'if it's what she wants.'

Ruth got as far as the doorway. Then she turned back.

'He doesn't care whether I'm tired or not,' Ruth said. 'He doesn't think of me any more. He only ever thinks about Mary Fisher; you know, the writer. She's his mistress.'

Bobbo was shocked by this indiscretion, this disloyalty, but also gratified. Ruth was not to be trusted. He'd always known it.

'Ruth,' he said, 'it's very unfair to my parents to involve them in our family problems. It's nothing to do with them. Have pity, will you, for once, on the helpless bystanders.'

'But it *is* something to do with me,' said Brenda. 'Your father never behaved like that; I don't know where you get it from.'

'Kindly respect my privacy, mother,' said Bobbo. 'It's the least you can do after the childhood I led.'

'And what was the matter with your childhood?' demanded Brenda, turning quite pink.

'Your mother's right,' said Angus. 'I think you should apologise to her for that. But fair's fair, Brenda, I think you should leave the young people to sort things out in their own way.'

'Father,' said Bobbo, 'it was just that kind of attitude in you that gave me one of the most appalling childhoods any child could have.'

Mary Fisher had lately been explaining the roots of his unhappiness to him.

'I never made your mother unhappy,' said Angus. 'Say what you like about me, but I never deliberately did harm to any woman.'

'Then all I can say is,' said Brenda, 'you did it by accident.'

'Women are always imagining things,' said Angus.

'Especially Ruth,' said Bobbo. 'Mary Fisher is one of my best clients. I'm very lucky to have her on my books. I certainly value her both as a creative person—she's remarkably talented—and I like to think as a friend, but I'm afraid our Ruth has a suspicious mind!'

Ruth looked from one parent-in-law to the other and then at her husband and dropped the tureen of mushroom soup, which flowed over the metal rim where the tiles stopped and the carpet began, and the children and the animals returned, summonsed by the sound of new disaster. Ruth thought that Harness was laughing.

'Perhaps Ruth ought to get out and get a job,' said Angus, on his knees on the floor, spooning soup back into a bowl, but less fast than the carpet absorbed it, so that he had to press the spoon hard into the pile to extract the precious grey liquid. 'Keep herself busy: less prone to imagining things.'

'There *are* no jobs,' Ruth pointed out.

'Nonsense,' said Angus. 'Anyone who really wants one can get one.'

'That's not true,' said Brenda. 'What with inflation, recession and so on . . . You don't mean us to *eat* that, do you, Angus?'

'Waste not, want not,' said Angus.

Bobbo wished to be far, far away, with Mary Fisher, to hear her bubbly laugh, hold her pale hand and put her little fingers one by one into his mouth until her breathing quickened and she wet her own lips with her pink, pink tongue.

Nicola kicked the cat, whose name was Mercy, out of the way, and the cat went straight to the grate and squatted, crapping its revenge, and Brenda wailed and pointed at Mercy, and Harness became over-excited and leapt up against Andy in semi-sexual assault, and Ruth just stood there, a giantess, and did nothing, and Bobbo lost his temper.

'See how I have to live!' He shouted. 'It's always like this. My wife creates havoc and destruction all round: she destroys everyone's happiness!'

'Why won't you love me?' wailed Ruth.

'How can one love,' shouted Bobbo, 'what is essentially unlovable?'

'You're both upset,' said Angus, giving up the soup to the carpet. 'You've been working too hard.'

'It's a lot for a woman,' said Brenda. 'Two growing children! And you were never easy, even as a boy, Bobbo.'

'I was perfectly easy,' yelled Bobbo. 'You just resented every moment you spent on me.'

'Come along, Brenda,' said Angus. 'Least said, soonest mended. We'll eat out.'

'A good idea,' shouted Bobbo, 'since my wife has already thrown your main course on the floor.'

'Temper, temper,' said Brenda. 'In Los Angeles they build houses without kitchens, because nobody bothers to cook. And quite right too.'

'But I spent all day doing this,' sobbed Ruth. 'And now no one's going to eat it.'

'Because it's uneatable!' shouted Bobbo. 'Why am I always surrounded by women who can't cook?'

'I'll ring you in the morning, pet,' said Brenda to Ruth. 'You have a nice bath and get a good night's sleep. You'll feel better then.'

'I shall never forgive you for being so rude to my mother,' said Bobbo to Ruth, coldly, and loud enough for his mother to hear.

'Don't you go putting the blame on her,' said Brenda, cunningly. 'It was you who was rude, not her. I am a perfectly good cook, I just don't care to do it.'

'Marriage isn't easy,' remarked Angus, putting on his coat. 'It's like parenthood, something people have to work at. Of course, usually it's left to one partner rather than the other.'

'It certainly is!' said Brenda meaningfully, drawing on her gloves. She was not focusing properly: she had forgotten to put anti-perspirant under her right arm, and her pretty tan blouse was beginning to show a single dark under-arm stain. She had a lop-sided look.

'Now do you see what's happening?' Bobbo turned on Ruth. 'You've even set my parents quarrelling! If you see happiness you have to destroy it. It's the kind of woman you are.'

Brenda and Angus left. They walked away down the path, side by side but not touching. Domestic strife is catching. Happy couples do well to avoid the company of the unhappy.

Ruth went into the bathroom and locked the door. Andy and Nicola took the chocolate mousse from the fridge and shared it.

'It would serve you right if I went to see Mary,' said Bobbo to Ruth, through the keyhole. 'You have worked terrible mischief here tonight! You have upset my parents, you have upset your children, and you have upset me. Even the animals were affected. I see you at last as you really are. You are a third-rate person. You are a bad mother, a worse wife and a dreadful cook. In fact I don't think you are a woman at all. I think that what you are is a she-devil!'

It seemed to him, when he said this, that there was a change in the texture of the silence that came from the other side of the door; he thought perhaps he had shocked her into submission and apology: but though he knocked and banged she still did not come out.

FAY WELDON, *The Life and Loves of a She-Devil*, 1983

Aunt Edith was a handsome woman with straight thick brows which remained raven black even when her hair had turned swan's-wing white, and the bitterest mouth that I have ever seen on anybody. She, alas for them both, had married another Archie, weak-willed and amiable, who did not tell her beforehand that he was a quarter Indian—his mother being the product of an Indian Army colonel and a rajah's daughter—what would have happened if he had told Aunt Edith before it was too late, there's no knowing. Maybe she would still have married him, but I very much doubt it. As it was, finding out afterwards, she refused to have children—I very much doubt if she even allowed him into her bed!—and set out to make his life a cold hell to his dying day. I have been there at some family gathering myself, puzzled as a dog may be by stresses in the air, the electric discharge of things I did not understand, when he came into the room, and Aunt Edith sniffed loudly and said, 'There's a most peculiar smell in this room. One would almost think that somebody black had come into it.'

For many years, the family were quite seriously prepared for Uncle Archie to murder her one day, and prepared, if he did, to go into the witness-box on his behalf and swear that he did it under unendurable provocation.

ROSEMARY SUTCLIFF, *Blue Remembered Hills*, 1983

As wife of the Archbishop of Canterbury, [Lady Fisher] once addressed the young girls of Wycombe Abbey School on the matter of marriage. During question time a thoughtful-looking pupil put her hand up to ask: 'Mrs Fisher, have you ever thought of divorce?' She replied: 'Of divorce, never; of murder, frequently.'

from *The Times*, October 1986

MRS CAUDLE HAS TAKEN COLD. THE TRAGEDY OF THIN SHOES.

'No, Caudle; I wouldn't wish to say anything to accuse you; no, goodness knows, I wouldn't make you uncomfortable for the world,—but the cold I've got, I got ten years ago. I have never said anything about it—but it has never left me. Yes; ten years ago the day before yesterday. *How can I recollect it?* Oh, very well: women remember things you never think of: poor souls! they've good cause to do so. Ten years ago, I was sitting up for you,—there now, I'm not going to say anything to vex you, only do let me speak: ten years ago, I was waiting for you,—and I fell asleep, and the fire went out, and when I woke I found I was sitting right in the draught of the key-hole. That was my death, Caudle, though don't let that make you uneasy, love; for I don't think you meant to do it.

'Ha! it's all very well for you to call it nonsense; and to lay your ill conduct upon my shoes. That's like a man, exactly! There never was a man yet that killed his wife, who couldn't give a good reason for it. No: I don't mean to say that you've killed me: quite the reverse: still, there's never been a day that I haven't felt that key-hole. What? *Why won't I have a doctor?* What's the use of a doctor? Why should I put you to expense? Besides, I dare say you'll do very well without me, Caudle: yes, after a very little time, you won't miss me much—no man ever does.

'Peggy tells me, Miss Prettyman called to-day. *What of it?* Nothing, of course. Yes; I know she heard I was ill, and that's why she came. A little indecent, I think, Mr Caudle; she might wait; I sha'n't be in her way long; she may soon have the key of the caddy, now.

'Ha! Mr Caudle, what's the use of your calling me your dearest soul now? Well, I do believe you. I dare say you do mean it; that is, I hope you do. Nevertheless, you can't expect I can lie quiet in this bed, and think of that young woman—not, indeed, that she's near so young as she gives herself out. I bear no malice towards her, Caudle,—not the least. Still, I don't think I could lay at peace in my grave if—well, I won't say anything more about her; but you know what I mean.

'I think dear mother would keep house beautifully for you, when I'm gone. Well, love, I won't talk in that way if you desire it. Still, I know I've a dreadful cold; though I won't allow it for a minute to be the shoes—certainly not. I never would wear 'em thick, and you know it, and they never gave me cold yet. No, dearest Caudle, it's ten years ago that did it; not that I'll say a syllable of the matter to hurt you. I'd die first.

'Mother, you see, knows all your little ways; and you wouldn't get another wife to study you and pet you up as I've done—a second wife never does; it isn't likely she should. And after all, we've been very happy. It hasn't been my fault, if we've ever had a word or two, for you couldn't help now and then being aggravating; nobody can help their tempers always—especially men. Still we've been very happy, haven't we, Caudle?

'Good night. Yes,—this cold does tear me to pieces; but for all that, it isn't the shoes. God bless you, Caudle; no—it's *not* the shoes. I won't say it's the key-hole; but again I say, it's not the shoes. God bless you once more—But never say it's the shoes.'

DOUGLAS JERROLD, *Mrs Caudle's Curtain Lectures*, 1846

1895

Since I felt that my only crime was wanting to copy it [Tolstoy's manuscript] out, it suddenly flashed across my mind that there must be some far more serious reason why he should want to leave me, and that this was just an excuse. I immediately thought of that woman and lost all my self-control, and so as he should not be the one to leave me, I ran out of the house and tore off down the road. He came chasing after me, I in my dressing-gown, he in his pants and waistcoat, without a shirt. He pleaded with me to go back, but at that point I had only one wish and that was to die, never mind how. I remember I was sobbing and shouting: 'I don't care, let them take me away and put me in prison or the mental hospital!' Lyovochka dragged me back to the house, I kept falling in the snow and got soaked to the skin. I had only a night-dress on under my dressing-gown, and had nothing on my feet but a pair of slippers, and I am now ill, demented and choked, and cannot think at all clearly. . . .

Feelings of jealousy and rage, the mortifying thought that he *never did anything for me*, the old grief of having loved him so much when he had never loved me—all this reduced me to a state of utter despair. I flung the proofs on the table, threw on a light overcoat, put on my galoshes and hat, and slipped out of the house. Unfortunately—or perhaps fortunately—Masha [her daughter] had noticed my distraught face and followed me, although I didn't realise this at the time. I stumbled towards the Convent of the Virgin, intending to freeze to death in a wood on the Sparrow Hills. . . . I didn't regret what I was doing. I had staked almost my whole life on one card—my love for my husband—and now the game was lost and I had nothing to live for. I wasn't sorry about the children either. I always feel that however much *we* may love them, *they* never love *us*, and I was sure they would survive quite well without me. Masha, as it turned out, had not let me out of her sight, and she eventually managed to take me home. But my despair did not subside, and for two days I kept trying to leave. The following day I just hailed a cab in the street and set off for the Kursk station. How my children guessed that I had gone there I shall never know. But Seryozha and Masha caught up with me and once again took me home. Each time I got back I felt so foolish and ashamed of myself. That night (February 7) I was very ill. All the painful feelings inside me intensified to a point of unbearable anguish. I vaguely recall thinking that anybody Lyovochka touched was doomed to perish. . . . I still feel that my love for him will be the end of me, and will destroy my soul. If I do manage to release myself from it—from loving him I mean—I shall be saved. If not, one way or another, I am done for. He has killed my very soul—I am already dead!

After I'd been sobbing for a long time he came in, kneeled before me on the floor and begged me to forgive him. If he could keep just a fragment of that compassion for me alive, then I might still be happy with him.

Having tortured my soul, he then called in the doctors to examine me.

*

1910

By that evening I was feeling very bad indeed, and the spasms in my heart, my aching head and unbearable feelings of despair were making me shudder all over; my teeth were chattering, I was choking and sobbing, I thought I was dying. I cannot remember ever in my whole life being in such a frightful emotional state. I was terrified, and in a desperate attempt to save myself I naturally threw myself at the mercy of the man I love, and I myself sent him a second telegram: 'Implore you come tomorrow, 23rd.' But on the morning of the 23rd, instead of taking the 11 a.m. train and coming to my help, he sent a telegram which said: '*More convenient* return morning 24th. If necessary, will take night train.' . . .

On the evening of the 23rd Lev Nik. returned—with his hangers-on—in a disgruntled and unfriendly mood. . . .

We had a painful talk, and I said everything that was on my mind. Lev Nik. sat on a stool looking hunched and wretched, and said almost nothing. What could he have said? There were moments when I felt dreadfully sorry for him. The only reason I didn't poison myself that time is that I am such a *coward*. But there were good reasons for doing so, and I hope the Lord will take me free of a sinful suicide.

While we were having our painful talk a wild beast suddenly leapt out of Lev Nik., his eyes blazed with rage and he said something so cutting that at that moment I hated him and said: 'Ah, so that is what you are really like!' And he grew quiet immediately.

The next morning my undying love for him got the better of me, and when he came into the room I threw myself into his arms asking him to forgive me, take pity on me and be nice to me; he embraced me and wept, and we both decided that henceforth everything would be different, and we would love and cherish one another. I wonder how long this will last.

But then I couldn't bear to tear myself away from him; I wanted to be close to him, to be one with him. I asked him to go to Ovsyannikovo with me, as I wanted to be with him, and he made a great effort to please me and agreed to go, although he obviously didn't want to. Then on the road he kept trying to get away from me and getting out and walking, and I began weeping again, as there seemed no point in going if I was just going to drive along in the carriage on my own.

But we arrived together and I became calmer—there was some small ray of happiness in just being *together*. . . .

Whatever I may have done, no one could possibly have given my husband more than I have. I have loved him passionately, selflessly, honestly and considerately, I have surrounded him with care, I have guarded him, I have helped him as best I could, I have not betrayed him with so much as a word or a gesture—what more can a woman give than this most intense love? I am 16 years younger than my husband, and have always looked 10 years younger than my age, but I have given my passion, love, health and energy to him and to no one else. I now realise that my husband's sacred philosophy of life is only in books; he needs a comfortable, settled existence for his work, and has lived all his life in these conditions—as though it were for *my* sake! God be with him, and Lord help me! Help people to discover the *truth* and see through this Phariseeism! For whatever they may plot against me, Lev N.'s love for me pervades everything he does, and everyone must ask themselves: if two people live together and love one another for 48 years, must there not have been *some reason* for this love?

SOFIA TOLSTOY, *Diaries*

¶ *The Tolstoys' marriage continued with the same passionate intensity, 'undying love' on her part, cruel rejection on his, alternating with tearful reconciliations, until Tolstoy's death in the railway station at Astapova, to which he had fled, aged 82, from his wife and home.*

George and Martha's relationship is based on mutual disillusion and despair. Their vicious attacks and counter-attacks are the only form of intimacy they can tolerate, and their ritual battles are as binding, as excluding of others, and ultimately as tension-releasing as sexual intercourse, for which they serve as a substitute:

GEORGE: You can sit there in that chair of yours, you can sit there with the gin running out of your mouth, and you can humiliate me, you can tear me apart . . . ALL NIGHT . . . and that's perfectly all right . . . that's O.K. . . .

MARTHA: YOU CAN STAND IT!

GEORGE: I CANNOT STAND IT!

MARTHA: YOU CAN STAND IT! YOU MARRIED ME FOR IT!!

[*A silence.*]

GEORGE [*quietly*]: That is a desperately sick lie.

MARTHA: DON'T YOU KNOW IT, EVEN YET?

GEORGE [*shaking his head*]: Oh . . . Martha.

MARTHA: My arm has gotten tired whipping you.

GEORGE [*stares at her in disbelief*]: You're mad.

MARTHA: For twenty-three years!

GEORGE: You're deluded . . . Martha, you're deluded.

MARTHA: IT'S NOT WHAT I'VE WANTED!

GEORGE: I thought at least you were . . . on to yourself. I didn't know. I . . . didn't know.

MARTHA [*anger taking over*]: I'm on to myself.

GEORGE [*as if she were some sort of bug*]: No . . . no . . . you're . . . sick.

MARTHA [*rises—screams*]: I'LL SHOW YOU WHO'S SICK!

GEORGE: All right, Martha . . . you're going too far.

MARTHA [*screams again*]: I'LL SHOW YOU WHO'S SICK. I'LL SHOW YOU.

GEORGE [*he shakes her*]: Stop it! [*Pushes her back in her chair.*] Now, stop it!

MARTHA [*calmer*]: I'll show you who's sick. [*Calmer*] Boy, you're really having a field day, hunh? Well, I'm going to finish you . . . before I'm through with you. . . . before I'm through with you you'll wish you'd died in that automobile, you bastard.

GEORGE [*emphasizing with his forefinger*]: And you'll wish you'd never mentioned our son!

MARTHA [*dripping contempt*]: You . . .

GEORGE: Now, I said I warned you.

MARTHA: I'm impressed.

GEORGE: I warned you not to go too far.

MARTHA: I'm just beginning.

GEORGE [*calmly, matter-of-factly*]: I'm numbed enough . . . and I don't mean by liquor, though maybe that's been part of the process—a gradual, over-the-years going to sleep of the brain cells—I'm numbed enough, now, to be able to take you when we're alone. I don't listen to you . . . or when I *do* listen to you, I sift everything, I bring everything down to reflex response, so I don't really *hear* you, which is the only way to manage it. But you've taken a new tack, Martha, over the past couple of centuries—or however long it's been I've lived in this house with you—that makes it just too much . . . too much. I don't mind your dirty underthings in public . . . well, I *do* mind, but I've reconciled myself to that . . . but you've moved bag and baggage into your own fantasy world now, and you've started playing variations on your own distortions, and, as a result . . .

MARTHA: Nuts!

GEORGE: Yes . . . you have.

MARTHA: Nuts!

GEORGE: Well, you can go on like that as long as you want to. And, when you're done . . .

MARTHA: Have you ever listened to your sentences, George? Have you ever listened to the way you talk? You're so frigging . . . convoluted . . . that's what you are. You talk like you were writing one of your stupid papers.

GEORGE: Actually, I'm rather worried about you. About your mind.

MARTHA: Don't you worry about my mind, sweetheart!

GEORGE: I think I'll have you committed.

MARTHA: You WHAT?

GEORGE [*quietly . . . distinctly*]: I think I'll have you committed.

MARTHA [*breaks into long laughter*]: Oh baby, aren't you something!

GEORGE: I've got to find some way to really get at you.

MARTHA: You've got at me, George . . . you don't have to do anything. Twenty-three years of you has been quite enough.

GEORGE: Will you go quietly, then?

MARTHA: You know what's happened, George? You want to know what's *really happened*? [*Snaps her fingers.*] It's snapped, finally. Not me . . . *it*. The whole arrangement. You can go along . . . forever, and everything's . . . manageable. You make all sorts of excuses to yourself . . . *you* know . . . this is life . . . the hell with it . . . maybe tomorrow he'll be dead . . . maybe tomorrow *you'll* be dead . . . all sorts of excuses. But then, one day, one night, something happens . . . and SNAP! It breaks. And you just don't give a damn any more. I've tried with you, baby . . . really, I've tried.

GEORGE: Come off it, Martha.

MARTHA: I've tried . . . I've really tried.

GEORGE [*with some awe*]: You're a monster . . . you *are*.

MARTHA: I'm loud, and I'm vulgar, and I wear the pants in this house because somebody's got to, but I am *not* a monster. I am *not*.

GEORGE: You're a spoiled, self-indulgent, wilful, dirty-minded, liquor-ridden . . .

MARTHA: SNAP! It went snap. Look, I'm not going to try to get through to you any more. . . . I'm not going to try. There was a second back there, maybe, there was a second, just a second, when I could have gotten through to you, when maybe we could have cut through all this crap. But that's past, and now I'm not going to try.

GEORGE: Once a month, Martha! I've gotten used to it . . . once a month and we get misunderstood Martha, the good-hearted girl underneath the barnacles, the little Miss that the touch of kindness'd bring to bloom again. And I've believed it more times than I want to remember, because I don't want to think I'm that much of a sucker. I don't believe you ... I just don't believe you. There is no moment ... there is no moment any more when we could ... come together.

MARTHA [*armed again*]: Well, maybe you're right, baby. You can't come together with nothing, and you're nothing! SNAP! It went snap tonight at Daddy's party. [*Dripping contempt, but there is fury and loss under it.*] I sat there at Daddy's party, and I watched you ... I watched you sitting there, and I watched the younger men around you, the men who were going to go somewhere. And I sat there and I watched you, and *you* weren't *there*! And it snapped! It finally snapped! And I'm going to howl it out, and I'm not going to give a damn what I do, and I'm going to make the damned biggest explosion you ever heard.

GEORGE [*very pointedly*]: You try it and I'll beat you at your own game.

MARTHA [*hopefully*]: Is that a threat, George? Hunh?

GEORGE: That's a threat, Martha.

MARTHA [*fake-spits at him*]: You're going to get it, baby.

GEORGE: Be careful, Martha ... I'll rip you to pieces.

MARTHA: You aren't man enough ... you haven't got the guts.

GEORGE: Total war?

MARTHA: Total.

[*Silence. They both seem relieved ... elated.*

EDWARD ALBEE, *Who's Afraid of Virginia Woolf*, 1962

EPISODE OF DECAY

Being very religious, she devoted most of her time to fear.
Under her calm visage, terror held her,
Terror of water, of air, of earth, of thought,
Terror lest she be disturbed in her routine of eating her husband.
She fattened on his decay, but she let him decay without pain.
And still she would ask, while she consumed him particle by particle,
'Do you wish me to take it, dear? Will it make you happier?'
And down the plump throat he went day after day in tidbits;
And he mistook the drain for happiness,
Could hardly live without the deadly nibbling ...

She had eaten away the core of him under the shell,
Eaten his heart and drunk away his breath;
Till on Saturday, the seventeenth of April,
She made her breakfast on an edge of his mind.
He was very quiet that day, without knowing why.
A last valiant cell of his mind may have been insisting that the fault was not hers but his;
But soon he resumed a numbness of content;
The little cell may have been thinking that one dies sooner or later
And that one's death may as well be useful. . . .
For supper, he offered her tea and cake from behind his left ear;
And after supper they took together the walk they always took together after supper.

WITTER BYNNER

•

HICKEY (*in a tone of fond, sentimental reminiscence*): Yes, sir, as far back as I can remember, Evelyn and I loved each other. She always stuck up for me. She wouldn't believe the gossip—or she'd pretend she didn't. No one could convince her I was no good. Evelyn was stubborn as all hell once she'd made up her mind. Even when I'd admit things and ask her forgiveness, she'd make excuses for me and defend me against myself. She'd kiss me and say she knew I didn't mean it and I wouldn't do it again. So I'd promise I wouldn't. I'd have to promise, she was so sweet and good, though I knew darned well— (*A touch of strange bitterness comes into his voice for a moment.*) No, sir, you couldn't stop Evelyn. Nothing on earth could shake her faith in me. Even I couldn't. She was a sucker for a pipe dream. (*Then quickly.*) Well, naturally, her family forbid her seeing me. They were one of the town's best, rich for that hick burg, owned the trolley line and lumber company. Strict Methodists, too. They hated my guts. But they couldn't stop Evelyn. She'd sneak notes to me and meet me on the sly. I was getting more restless. The town was getting more like a jail. I made up my mind to beat it. . . . The night before I left town, I had a date with Evelyn. I got all worked up, she was so pretty and sweet and good. I told her straight, 'You better forget me, Evelyn, for your own sake. I'm no good and never will be. I'm not worthy to wipe your shoes.' I broke down and cried. She just said, looking white and scared, 'Why, Teddy? Don't you still love me?' I said, 'Love you? God, Evelyn, I love you more than anything in the world. And I always will!' She said, 'Then nothing else matters, Teddy, because nothing but death could stop my loving you. So I'll wait, and when you're ready you send for me and we'll be married. I know I can make you happy, Teddy, and once you're happy you won't want to do any of the bad things you've done any more.' And I said, 'Of course, I won't, Evelyn!' I meant it, too. I believed it. I loved her so much she could make me believe anything. (*He sighs.*) . . . So I beat it to the Big Town. I got a job easy, and it was a cinch for me to make good. I had the knack. . . .

It was fun. But still, all the while I felt guilty, as if I had no right to be having such a good time away from Evelyn. In each letter I'd tell her how I missed her, but I'd keep warning her, too. I'd tell her all my faults, how I liked my booze every once in a while, and so on. But there was no shaking Evelyn's belief in me, or her dreams about the future. After each letter of hers, I'd be as full of faith as she was. So as soon as I got enough saved to start us off, I sent for her and we got married. Christ, wasn't I happy for a while! And wasn't she happy! I don't care what anyone says, I'll bet there never was two people who loved each other more than me and Evelyn. Not only then but always after, in spite of everything I did— (*He pauses—then sadly.*) Well, it's all there, at the start, everything that happened afterwards. I never could learn to handle temptation. I'd want to reform and mean it. I'd promise Evelyn, and I'd promise myself, and I'd believe it. I'd tell her, it's the last time. And she'd say, 'I know it's the last time, Teddy. You'll never do it again.' That's what made it so hard. That's what made me feel such a rotten skunk—her always forgiving me. My playing around with women, for instance. It was only a harmless good time to me. Didn't mean anything. But I'd know what it meant to Evelyn. So I'd say to myself, never again. But you know how it is, travelling around. The damned hotel rooms. I'd get seeing things in the wall paper. I'd get bored as hell. Lonely and homesick. But at the same time sick of home. I'd feel free and I'd want to celebrate a little. I never drank on the job, so it had to be dames. Any tart. What I'd want was some tramp I could be myself with without being ashamed—someone I could tell a dirty joke to and she'd laugh. . . .

Sometimes I'd try some joke I thought was a corker on Evelyn. She'd always make herself laugh. But I could tell she thought it was dirty, not funny. And Evelyn always knew about the tarts I'd been with when I came home from a trip. She'd kiss me and look in my eyes, and she'd know. I'd see in her eyes how she was trying not to know, and then telling herself even if it was true, he couldn't help it, they tempt him, and he's lonely, he hasn't got me, it's only his body, anyway, he doesn't love them, I'm the only one he loves. She was right, too. I never loved anyone else. Couldn't if I wanted to. (*He pauses.*) She forgave me even when it all had to come out in the open. You know how it is when you keep taking chances. You may be lucky for a long time, but you get nicked in the end. I picked up a nail from some tart in Altoona. . . .

I had to do a lot of lying and stalling when I got home. It didn't do any good. The quack I went to got all my dough and then told me I was cured and I took his word. But I wasn't, and poor Evelyn— But she did her best to make me believe she fell for my lie about how travelling men get things from drinking-cups on trains. Anyway, she forgave me. The same way she forgave me every time I'd turn up after a periodical drunk. You all know what I'd be like at the end of one. You've seen me. Like something lying in the gutter that no alley cat would lower itself to drag in—something they threw out of the D.T. ward in Bellevue along with the garbage, something that ought to be dead and isn't! (*His face is convulsed with self-loathing.*) Evelyn wouldn't have heard from me in a month or more. She'd have been waiting there alone, with the neighbours shaking their heads and feeling sorry for her out loud. That was before she got me to move to the outskirts, where there weren't any next-door neighbours. And then the door would open and in I'd stumble—looking like what I've said—into her home, where she kept everything so spotless and clean. And I'd sworn it would never happen again, and now I'd have to start swearing again this was the last time. I could see disgust having a battle in her eyes with love. Love always won. She'd make herself kiss me, as if nothing had happened, as if I'd just come home from a business trip. She'd never complain or bawl me out. (*He bursts out in a tone of anguish that has anger and hatred beneath it.*) Christ, can you imagine what a guilty skunk she made me feel! If she'd only admitted once she didn't believe any more in her pipe dream that some day I'd behave! But she never would. Evelyn was stubborn as hell. Once she'd set her heart on anything, you couldn't shake her faith that it had to come true—tomorrow! It was the same old story, over and over, for years and years. It kept piling up, inside her and inside me. God, can you picture all I made her suffer, and all the guilt she made me feel, and how I hated myself! If she only hadn't been so damned good—if she'd been the same kind of wife I was a husband. God, I used to pray sometimes she'd—I'd even say to her, 'Go on, why don't you, Evelyn? It'd serve me right. I wouldn't mind. I'd forgive you.' Of course, I'd pretend I was kidding—the same way I used to joke here about her being in the hay with the iceman. She'd have been so hurt if I'd said it seriously. She'd have thought I'd stopped loving her. (*He pauses—then looking around at them.*) I suppose you think I'm a liar, that no woman could have stood all she stood and still loved me so much—that it isn't human for any woman to be so pitying and forgiving. Well, I'm not lying, and if you'd ever seen her, you'd realize I wasn't. It was written all over her face, sweetness and love and pity and forgiveness. (*He reaches mechanically for the inside pocket of his coat.*) Wait! I'll show you. I always carry her picture. (*Suddenly he looks startled. He stares before him, his hand falling back—quietly.*) No, I'm forgetting I tore it up—afterwards. I didn't need it any more. (*He pauses.*) . . .

It kept piling up, like I've said. I got so I thought of it all the time. . . . It got so every night I'd wind up hiding my face in her lap, bawling and begging her forgiveness. And, of course, she'd always comfort me and say, 'Never mind, Teddy, I know you won't ever again.' Christ, I loved her so, but I began to hate that pipe dream! I began to be afraid I was going bughouse, because sometimes I couldn't forgive her for forgiving me. I even caught myself hating her for making me hate myself so much. There's a limit to the guilt you can feel and the forgiveness and the pity you can take! You have to begin blaming someone else, too. I got so sometimes when she'd kiss me it was like she did it on purpose to humiliate me, as if she'd spit in my face! But all the time I saw how crazy and rotten of me that was and it made me hate myself all the more. You'd never believe I could hate so much, a good-natured, happy-go-lucky slob like me. And as the time got nearer to when I was due to come here for my drunk around Harry's birthday, I got nearly crazy. I kept swearing to her every night that this time I really wouldn't, until I'd made it a real final test to myself—and to her. And she kept encouraging me and saying, 'I can see you really mean it now, Teddy. I know you'll conquer it this time, and we'll be so happy, dear.' . . . And I knew if I came this time, it was the finish. I'd never have the guts to go back and be forgiven again, and that would break Evelyn's heart because to her it would mean I didn't love her any more. (*He pauses.*) That last night, I'd driven myself crazy trying to figure some way out for her. I went in the bedroom. I was going to tell her it was the end. But I couldn't do that to her. She was sound asleep. I thought, God, if she'd only never wake up, she'd never know! And then it came to me—the only possible way out, for her sake. I remembered I'd given her a gun for protection while I was away and it was in the bureau drawer. She'd never feel any pain, never wake up from her dream. So I—

(*They all pound with their glasses and grumble in chorus:* 'Who the hell cares? We want to pass out in peace!'

The clamour of banging glasses dies out as abruptly as it started. Hickey hasn't appeared to hear it.)

HICKEY (*simply*): So I killed her.

(*There is a moment of dead silence.*) . . .

And then I saw I'd always known that was the only possible way to give her peace and free her from the misery of loving me. I saw it meant peace for me, too, knowing she was at peace. I felt as though a ton of guilt was lifted off my mind. I remember I stood by the bed and suddenly I had to laugh. I couldn't help it, and I knew Evelyn would forgive me. I remember I heard myself speaking to her, as if it was something I'd always wanted to say: 'Well, you know what you can do with your pipe dream now, you damned bitch!' (*He stops with a horrified start, as if shocked out of a nightmare, as if he couldn't believe he heard what he had just said. He stammers.*) No! I never—!

EUGENE O'NEILL, *The Iceman Cometh*, 1946

THE STRANGE CASE OF MR ORMANTUDE'S BRIDE

Once there was a bridegroom named Mr Ormantude whose
 intentions were hard to disparage,
Because he intended to make his a happy marriage,
And he succeeded for going on fifty years,
During which he was in marital bliss up to his ears.

His wife's days and nights were enjoyable
Because he catered to every foible;
He went around humming hymns
And anticipating her whims.
Many a fine bit of repartee died on his lips
Lest it throw her anecdotes into eclipse;
He was always silent when his cause was meritorious,
And he never engaged in argument unless sure he was so
obviously wrong that she couldn't help emerging victorious,
And always when in her vicinity
He was careful to make allowances for her femininity;
Were she snappish, he was sweetish,
And of understanding her he made a fetish.
Everybody said his chances of celebrating his golden wedding
looked good,
But on his golden wedding eve he was competently poisoned by
his wife who could no longer stand being perpetually
understood.

OGDEN NASH, 1943

CALL IT A GOOD MARRIAGE

Call it a good marriage—
For no one ever questioned
Her warmth, his masculinity,
Their interlocking views;
Except one stray graphologist
Who frowned in speculation
At her h's and her s's,
His p's and w's.

Though few would still subscribe
To the monogamic axiom
That strife below the hip-bones
Need not estrange the heart,
Call it a good marriage:
More drew those two together,
Despite a lack of children,
Than pulled them apart.

Call it a good marriage:
They never fought in public,
They acted circumspectly
And faced the world with pride;
Thus the hazards of their love-bed
Were none of our damned business—
Till as jurymen we sat upon
Two deaths by suicide.

ROBERT GRAVES, 1965

9

Putting Asunder

DIVORCE is almost as old an institution as marriage itself, or rather, as Voltaire would have it, marriage pre-dates divorce by only a few weeks. Yet old as divorce is—and it certainly existed at least three thousand years before the birth of Christ—the debate about the proper grounds for the dissolution of a marriage, about who should be allowed to divorce whom, and when, has never ceased.

Meanwhile the private pain, the agony of the decision, the profound loss of confidence and even of personal identity which the individual suffers through divorce, has not diminished.

Neither a sexual revolution nor enlightened legislation can eliminate overnight more than a thousand years of divorce taboo. It lives on like a ghost, chilly and half-perceived, generating a kind of moral shudder even in those to whom religion no longer means much and society does not accuse. Pride, or the superego, takes over where the Church and community leave off, and no amount of acceptance or understanding or social indifference lessens the sense of personal failure.

(A. Alvarez, *Life After Marriage*)

And Pharisees came up to him and tested him by asking, 'Is it lawful to divorce one's wife for any cause?' He answered, 'Have you not read that he who made them from the beginning made them male and female, and said, "For this reason a man shall leave his father and mother and be joined to his wife, and the two shall become one flesh"? So they are no longer two but one flesh. What therefore God has joined together, let not man put asunder.' They said to him, 'Why then did Moses command one to give a certificate of divorce, and to put her away?' He said to them, 'For your hardness of heart Moses allowed you to divorce your wives, but from the beginning it was not so. And I say to you: whoever divorces his wife, except for unchastity, and marries another, commits adultery.'

The disciples said to him, 'If such is the case of a man with his wife, it is not expedient to marry.' But he said to them, 'Not all men can receive this saying, but only those to whom it is given.'

Matthew 19: 3–11

As licentious as the world reputes me, I have (in good truth) more strictly observed the lawes of wedlock then either I had promised or hoped. It is no longer time to wince when one hath put on the shackles. A man ought wisely to husband his liberty, but after he hath once submitted himselfe into bondage, he is to stick unto it by the lawes of common duty, or at least enforce himselfe to keepe them. Those which undertake that covenant to deale therein with hate and contempt, do both injustly and incommodiously. If one do not alwaise discharge his duty, yet ought he at least ever love, ever acknowledge it. It is treason for one to marry unless he wed.

MONTAIGNE, *Essays* (trans. Florio), 1580

Marriage is the beginning and the pinnacle of all culture. It makes the savage gentle, and it gives the most cultivated the best occasion for demonstrating his gentleness. It has to be indissoluble: it brings so much happiness that individual instances of unhappiness do not come into account. And why speak of unhappiness at all? Impatience is what it really is, ever and again people are overcome by impatience, and then they like to think themselves unhappy. Let the moment pass, and you will count yourself happy that what has so long stood firm still stands. As for separation, there can be no adequate grounds for it. The human condition is compounded of so much joy and so much sorrow that it is impossible to reckon how much a husband owes a wife or a wife a husband. It is an infinite debt, it can be paid only in eternity. Marriage may sometimes be an uncomfortable state, I can well believe that, and that is as it should be. Are we not also married to our conscience, and would we not often like to be rid of it because it is more uncomfortable than a husband or a wife could ever be?

GOETHE, *Elective Affinities*, 1809

There is certainly a very strong and logical case to be made out for a marriage bond that is indissoluble even by death. It banishes step-parents from the world. It confers a dignity of tragic inevitability upon the association of husband and wife, and makes a love approach the gravest, most momentous thing in life. It banishes for ever any dream of escape from the presence and service of either party, or of any separation from the children of the union. It affords no alternative to 'making the best of it' for either husband or wife; they have taken a step as irrevocable as suicide. . . .

For my own part, I do not think the maintenance of a marriage that is indissoluble, that precludes the survivor from re-marriage, that gives neither party an external refuge from the misbehaviour of the other, and makes the children the absolute property of their parents until they grow up, would cause any very general unhappiness. Most people are reasonable enough, good-tempered enough, and adaptable enough to shake down even in a grip so rigid, and I would even go further and say that its very rigidity, the entire absence of any way out at all, would oblige innumerable people to accommodate themselves to its conditions and make a working success of unions that, under laxer conditions, would be almost certainly dissolved. . . .

A few more crimes of desperation perhaps might occur, to balance against an almost universal effort to achieve contentment and reconciliation. We should hear more of the 'natural law' permitting murder by the jealous husband or by the jealous wife, and the traffic in poisons would need a sedulous attention—but even there the impossibility of re-marriage would operate to restrain the impatient. On the whole, I can imagine the world rubbing along very well with marriage as unaccommodating as a perfected steel trap. . . .

And I take up this position because I believe in the family as the justification of marriage. Marriage to me is no mystical and eternal union, but a practical affair, to be judged as all practical things are judged—by its returns in happiness and human welfare. And directly we pass from the mists and glamours of amorous passion to the warm realities of the nursery, we pass into a new system of considerations altogether. . . .

Divorce as it exists at present is not a readjustment but a revenge. It is the nasty exposure of a private wrong. . . . Of course, if our divorce law exists mainly for the gratification of the fiercer sexual resentments, well and good, but if that is so, let us abandon our pretence that marriage is an institution for the establishment and protection of homes.

H. G. WELLS, *An Englishman Looks at the World*, 1914

Love is a punchy physical affair and therefore should not be confused with any other side of life or form of affection, and while it makes an agreeable foundation from which to begin a marriage the absence of physical love, *love* in fact, should never be allowed to interfere with the continuity of marriage. Marriage is the most important thing in life and must be kept going at almost any cost, it should only be embarked upon where there is, as well as physical love, a complete conformity of outlook. Women, as well as men, ought to have a great many love affairs before they marry as the most critical moment in a marriage is the falling off of physical love, which is bound to occur sooner or later and only an experienced woman can know how to cope with this. If not properly dealt with the marriage is bound to go on the rocks.

NANCY MITFORD, note in her appointments diary for 1941

Not even the intercourse of the sexes is exempt from the despotism of positive institution. Law pretends even to govern the indisciplinable wanderings of passion, to put fetters on the clearest deductions of reason, and by appeals to the will, to subdue the involuntary affections of our nature. Love is inevitably consequent upon the perception of loveliness. Love withers under constraint; its very essence is liberty; it is compatible neither with obedience, jealousy, nor fear; it is there most pure, perfect, and unlimited, where its votaries live in confidence, equality, and unreserve.

How long then ought the sexual connection to last? What law ought to specify the extent of the grievances which should limit its duration? A husband and wife ought to continue so long united as they love each other; any law which should bind them to cohabitation for one moment after the decay of their affection would be a most intolerable tyranny, and the most unworthy of toleration. How odious an usurpation of the right of private judgement should that law be considered which should make the ties of friendship indissoluble, in spite of the caprices, the inconsistency, the fallibility, and capacity for improvement of the human mind. And by so much would the fetters of love be heavier and more unendurable than those of friendship, as love is more vehement and capricious, more dependent on those delicate peculiarities of imagination, and less capable of reduction to the ostensible merits of the object.

*

But if happiness be the object of morality, of all human unions and disunions; if the worthiness of every action is to be estimated by the quantity of pleasurable sensation it is calculated to produce, then the connection of the sexes is so long sacred as it contributes to the comfort of the parties, and is naturally dissolved when its evils are greater than its benefits. There is nothing immoral in this separation. Constancy has nothing virtuous in itself, independently of the pleasure it confers, and partakes of the temporizing spirit of vice in proportion as it endures tamely moral defects of magnitude in the object of its indiscreet choice. Love is free; to promise for ever to love the same woman is not less absurd than to promise to believe the same creed; such a vow in both cases excludes us from all enquiry. The language of the votarist is this: the woman I now love may be infinitely inferior to many others; the creed I now profess may be a mass of errors and absurdities; but I exclude myself from all future information as to the amiability of the one and the truth of the other, resolving blindly and in spite of conviction to adhere to them: Is this the language of delicacy and reason? Is love of such a frigid heart of more worth than its belief?

PERCY BYSSHE SHELLEY (1792–1822), 'Against Legal Marriage'

Why is divorce unlawful but only for adultery? because, they say, that crime only breaks the matrimony. But this I reply, The institution itself gainsays: for that which is most contrary to the words and meaning of the institution, *that* most breaks the matrimony; but a perpetual unmeetness and unwillingness to all the duties of help, of love, and tranquillity, is most contrary to the words and meaning of the institution; that therefore much more breaks matrimony than the act of adultery, though repeated. For this, as it is not felt, nor troubles him that perceives it not, so being perceived, may be soon repented, soon amended: soon, if it can be pardoned, may be redeemed with the more ardent love and duty in her who hath the pardon. But this natural unmeetness both cannot be unknown long, and ever after cannot be amended, if it be natural; and will not, if it be far gone obstinate. So that wanting aught in the instant to be as great a breach as adultery, it gains it in the perpetuity to be greater. Next, adultery does not exclude her other fitness, her other pleasingness: she may otherwise be both loving and prevalent, as many adultresses be; but in this general unfitness or alienation she can be nothing to him that can please. In adultery nothing is given from the husband, which he misses, or enjoys the less, as it may be subtly given; but this unfitness defrauds him of the whole contentment which is sought in wedlock. And what benefit to him, though nothing be given by the stealth of adultery to another, if that which there is to give, whether it be solace, or society, be not such as may justly content him? and so not only deprives him of what it should give him, but gives him sorrow and affliction, which it did not owe him.

JOHN MILTON, *The Doctrine and Discipline of Divorce*, 1643

¶ *Milton married his first wife when he was 35 and she barely more than half his age. She shared none of his tastes and interests, having been brought up, in the words of John Aubrey, 'where there was a great deal of company, merriment, etc.'. She left him shortly after the wedding, and was only persuaded to return to him three years later. Milton did not divorce her on grounds of desertion, as he might have done, but became instead an early advocate of divorce on the grounds of incompatibility.*

And matrimony is there never broken but by death: except adultery break the bond, or else the intolerable wayward manners of either party. For if either of them find themselves for any such cause grieved, they may, by the license of the counsel, change and take another. But the other party liveth ever after in infamy, and out of wedlock. Howbeit, the husband to put away his wife for no other fault, but for that some mishap is fallen to her body, this by no means they will suffer! for they judge it a great point of cruelty, that any body in their most need of in that behalf have suffered wrong, being help and comfort should be cast of and forsaken; and that old age, which both bringeth sickness with it, and is a sickness itself, should unkindly and unfaithfully be dealt withal. But now and then it chanceth, whereas the man and woman cannot well agree between themselves, both of them finding other with whom they hope to live more quietly and merrily, that they, by the full consent of them both, be divorced asunder and married again to other. But that not without the authority of the counsel: which agreeth to no divorces before they and their wives have diligently tried and examined the matter. Yea, and then also they be loth to consent to it; because they know this to be the next way to break love between man and wife—to be in easy hope of a new marriage!

THOMAS MORE, *Utopia*, 1516, trans. Raphe Robinson, 1551

JUDGEMENT ON A BIGAMIST

'Prisoner at the bar, you have been convicted before me of what the law regards as a very grave and serious offence, that of going through the marriage ceremony a second time while your wife was still alive. You plead in mitigation of your conduct that she was given to dissipation and drunkenness, that she proved herself a curse to your household while she remained mistress of it, and that she had latterly deserted you; but I am not permitted to recognize any such plea. You had entered into a solemn engagement to take her for better, for worse, and if you got infinitely more of the latter, as you appear to have done, it was your duty patiently to submit. You say you took another person to be your wife because you were left with several young children, who required the care and protection of some one who might act as a substitute for the parent who had deserted them; but the law makes no allowances for bigamists with large families. Had you taken the other female to live with you as your concubine you would never have been interfered with by the law. But your crime consists in having—to use your own language—preferred to make an honest woman of her. Another of your irrational excuses is that your wife had committed adultery, and so you thought you were relieved from treating her with any further consideration; but you were mistaken. The law in its wisdom points out a means by which you might rid yourself from further association with a woman who had dishonoured you; but you did not think proper to adopt it. I will tell you what that process is. You ought first to have brought an action against your wife's seducer if you could discover him; that might have cost you money, and you say you are a poor working man, but that is not the fault of the law. You would then be obliged to prove by evidence your wife's criminality in a court of justice, and thus obtain a verdict with damages against the defendant, who was not unlikely to turn out to be a pauper. But so jealous is the law (which you ought to be aware is the perfection of reason) of the sanctity of the marriage tie, that in accomplishing all this you would only have fulfilled the lighter portions of your duty. You must then have gone, with your verdict in your hand, and petitioned the House of Lords for a divorce. It would cost you perhaps five or six hundred pounds, and you do not seem to be worth as many pence. But it is the boast of the law that it is impartial, and makes no difference between the rich and the poor. The wealthiest man in the kingdom would have had to pay no less than that sum for the same luxury; so that you would have no reason to complain. You would, of course, have to prove your case over again, and at the end of a year, or possibly two, you might obtain a decree which would enable you legally to do what you have thought proper to do without it. You have thus wilfully rejected the boon the legislature offered you, and it is my duty to pass upon you such sentence as I think your offence deserves, and that sentence is, that you be imprisoned for *one day*; and inasmuch as the present assizes is three days old, the result is that you will be immediately discharged.'

MR JUSTICE MAULE (Sir William Henry Maule) (1788–1858)

¶ *According to the* Biographical Dictionary of the Common Law, *Mr Justice Maule's ironical comments from the Bench were often misunderstood, despite his high reputation for 'subtlety, common sense and ability to free himself from the straitjacket of legal technicalities'.*

Why should such a foolish Marriage Vow
  Which long ago was made,
Oblige us to each other now
  When Passion is decay'd?
We lov'd, and we lov'd, as long as we cou'd,
  Till our Love was lov'd out in us both:
But our Marriage is dead, when the Pleasure is fled:
  'Twas Pleasure first made it an Oath.

If I have Pleasures for a Friend,
  And farther love in store,
What wrong has he whose joys did end,
  And who cou'd give no more?
'Tis a madness that he should be jealous of me,
  Or that I shou'd bar him of another:
For all we can gain, is to give our selves pain,
  When neither can hinder the other.

JOHN DRYDEN, from *Marriage à la Mode*, 1672

In the nature of the thing, the reasons . . . for a divorce are unnecessary; because, whatever causes the law may admit as sufficient to break a marriage, a mutual antipathy must be stronger than them all.

MONTESQUIEU, *The Spirit of Laws*, 1748

Why people divorce each other is their own business, inscrutable. They seldom, in my experience, know why themselves—know what was most important, I mean, along the camps of the Everest of dissatisfactions nearly any human being feels with any other human being he knows inside out. Maybe nothing is most important. It's the mountain, and you must get too weary to climb on.

JOHN BERRYMAN, *The Freedom of the Poet*, 1976

I have seen the failure of too many marriages, and warmed my hands before a few glowing successes, and in every case have thought and examined and wondered, and never found any logic, for marriage is as mysterious as life or fire. I know only that there is no objective existence to it. It cannot be studied from the outside and conclusions drawn, for it lives only inside itself, and what is presented to the observer or the listener bears only the relationship of a distorted shadow to the reality inside. There is no right and no wrong; there is no truth and no falsehood, neither wronged husband nor betrayed wife. There is only marriage, which is a mystical union, and when the partners to it cease to know that they are one flesh, when the mystical sense goes, there is no marriage. There is instead a practical problem, with a hundred factors pressing on the mind and the emotions. What to do? To pretend, to fight? To stay, to go? To deaden the senses or quicken them? To think of the children, who must be loved, or think of the mystery, which must be refound, without which there is no love? To die, to live?

JOHN MASTERS, *The Road Past Mandalay*, 1961

If she did not love you, she would not be
Jealous. I wish you would appreciate
A little more, how justified she is to hate.
You locked the door and threw away the key.
Don't try to tell me, that I cannot see
She does not own you. She's your chosen mate.
And you yourself should know the married state
Is not a contract, but a mystery.

I am not now the man I was before.
I'm sorry I don't like her any more.
What right have you to ask me to pretend
That all desire for her's not at an end?
The unexpected movements of one's soul
Arise completely out of one's control.

R. D. LAING, 1979

'I have heard it said that women love men even for their vices,' Anna began suddenly, 'but I hate him for his virtues. I can't live with him. Do you understand—the sight of him has a physical effect on me? It puts me beside myself. I can't, I can't live with him. What am I to do? I was unhappy before, and used to think one couldn't be more unhappy, but the awful state of things I am going through now, I could never have conceived. Would you believe it—knowing he is a good excellent man, that I am not worth his little finger, still I hate him! I hate him for his generosity. And there is nothing left for me but . . .'

She was going to say 'death', but Oblonsky would not let her finish.

'You are ill and overwrought,' he said. 'Believe me, you are exaggerating dreadfully. Nothing is so very terrible.'

*

'No, Stiva,' she said, 'I'm lost, lost! Worse than lost! I'm not lost yet—I can't say that all is over: on the contrary, I feel that it's not ended. I'm like an over-strained violin string that must snap. But it's not ended yet . . . and the end will be terrible.'

'Oh no, the string can be loosened by degrees. There is no situation from which there is no way out.'

'I have thought and thought. There is only one . . .'

Again he knew from her terror-stricken face that this one way of escape in her mind was death, and he would not let her say it.

'Not at all,' he said. 'Listen to me. You can't see your own position as I can. Let me give you my candid opinion.' Again he cautiously smiled his almond-oil smile. 'I'll begin from the beginning. You married a man twenty years older than yourself. You married him without love, or without knowing what love was. That was a mistake, let's admit it.'

'A fearful mistake!' said Anna.

'But I repeat—it's an accomplished fact. Then you had, let us say, the misfortune to fall in love with a man not your husband. That was a misfortune, but that, too, is an accomplished fact. And your husband accepted it and forgave it.' He stopped after each sentence, waiting for her to object, but she said nothing. 'That is how matters stand. Now the question is—can you go on living with your husband? Do you wish it? Does he wish it?'

'I don't know, I don't know at all.'

'But you said yourself that you can't endure him.'

'No, I didn't say so. I take it back. I don't know anything, I can't tell.'

'Yes, but let . . .'

'You can't understand. I feel I'm flying headlong over some precipice, but ought not to save myself. And I can't.'

'Never mind, we'll hold something out and catch you. I understand you, understand that you can't take it on yourself to express your wishes, your feelings.'

'There's nothing, nothing I wish . . . except for it to be all over.'

'But he sees that and knows it. Do you really suppose it weighs on him any less than on you? You're wretched, he's wretched, and what good can come of it? While a divorce would solve everything,' Oblonsky got out at last, not without difficulty expressing his central idea, and looked at her significantly.

She made no reply, and shook her cropped head in dissent. But from the look on her face, suddenly illuminated with its old beauty, he saw that if she did not desire this it was because it seemed to her unattainable happiness.

*

'If you care to know my opinion,' began Oblonsky with the same soothing smile of almond-oil tenderness with which he had addressed Anna. His kindly smile was so persuasive that Karenin, conscious of his own weakness and yielding to it, was involuntarily prepared to accept what Oblonsky should say. 'She would never say so, but there is one way out, one thing she might desire,' Oblonsky went on, 'and that is to terminate your relations and everything that reminds her of them. To my way of thinking, what's essential in your case is to get on a new basis with each other. And that can only be done by both sides having their freedom.'

'Divorce,' Karenin interrupted with aversion.

'Yes, I imagine that divorce—yes, divorce,' repeated Oblonsky, reddening. 'From every point of view that is the most sensible solution for a couple who find themselves in the position you are in. What else can they do if they find life impossible together? It is a thing that may always happen.'

Karenin sighed heavily and closed his eyes.

'There is only one point to be considered: does either party wish to marry again? If not, it is very simple,' said Oblonsky, by degrees losing his embarrassment.

Karenin, his face drawn with emotion, muttered something to himself and made no reply. What appeared so simple to Oblonsky, he had thought over thousands and thousands of times, and, far from being simple, it all seemed to him utterly impossible. An action for divorce, with the details of which he was now acquainted, appeared to him out of the question, because his feelings of self-respect and his regard for religion forbade his pleading guilty to a fictitious act of adultery, and still less could he allow his wife, forgiven and beloved by him, to be exposed and put to shame. Divorce seemed to him impossible also on other still more weighty grounds.

In the event of a divorce, what would become of his son? To leave him with his mother was out of the question. The divorced mother would have her own illegitimate family, in which the position and upbringing of a stepson would in all probability be wretched. Should he keep the child himself? He knew that would be an act of vengeance on his part, and he did not want that. But apart from this, what ruled out divorce more than anything in his eyes was that by consenting to a divorce he would be handing Anna over to destruction. . . . To consent to a divorce, to give her her freedom, would mean, as he saw it, to deprive himself of the last tie that bound him to life—the children he loved; and to take from her the last prop that supported her on the path of virtue, to thrust her to perdition. . . .

'Oh God, oh God! How have I deserved this?' thought Karenin . . .

LEO TOLSTOY, *Anna Karenina*, 1873

In my humble opinion marriage should be encouraged in every way, and divorce should be encouraged, not for its own sake, but for the sake of marriage. Regarded as a human institution, marriage cannot hope to be a working success unless divorce is in the background as a reserve. With divorce as a protection against unforeseen calamities arising out of marriage, marriage becomes a wise investment having regard to the circumstances generally. Without divorce I look upon marriage as a dangerous, mad gamble.

I look upon divorce as a policy of insurance. I take it it is a fact there is no marriage, however judiciously and carefully it may be arranged, whatever may be the absolute good faith of the parties to it, which is not an experiment. You cannot prevent it being anything else, and I therefore look upon divorce as a policy of insurance, providing an opportunity of relief and release to married couples who, through no fault of their own, without any moral blame, have come into contact with unforeseen difficulties and calamities which make married life intolerable.

Witness before the Royal Commission on Divorce, 1909

Divorce, in fact, is not the destruction of marriage, but the first condition of its maintenance.

GEORGE BERNARD SHAW (1856–1950)

MRS SULLEN: Spouse.
SULLEN: Ribb.
MRS SULLEN: How long have we been marry'd?
SULLEN: By the Almanack fourteen Months—But by my Account fourteen Years.
MRS SULLEN: 'Tis thereabout by my reckoning. . . . Pray, Spouse, what did you marry for?
SULLEN: To get an Heir to my Estate.
SIR CHARLES: And have you succeeded?
SULLEN: No.
ARCHER: The Condition fails of his side—Pray, Madam, what did you marry for?
MRS SULLEN: To support the Weakness of my Sex by the Strength of his, and to enjoy the Pleasures of an agreeable Society.
SIR CHARLES: Are your Expectations answer'd?
MRS SULLEN: No. . . .
SIR CHARLES: What are the Bars to your mutual Contentment.
MRS SULLEN: In the first Place I can't drink Ale with him.
SULLEN: Nor can I drink Tea with her.
MRS SULLEN: I can't hunt with you.
SULLEN: Nor can I dance with you.
MRS SULLEN: I hate Cocking and Racing.
SULLEN: And I abhor Ombre and Piquet.
MRS SULLEN: Your Silence is intollerable.
SULLEN: Your Prating is worse.
MRS SULLEN: Have we not been a perpetual Offence to each other—A gnawing Vulture at the Heart.
SULLEN: A frightful Goblin to the Sight.
MRS SULLEN: A Porcupine to the Feeling.
SULLEN: Perpetual Wormwood to the Taste.
MRS SULLEN: Is there on Earth a thing we cou'd agree in?
SULLEN: Yes—To part.
MRS SULLEN: With all my Heart.
SULLEN: Your Hand.
MRS SULLEN: Here.

SULLEN: These Hands join'd us, these shall part us—away—
MRS SULLEN: North.
SULLEN: South.
MRS SULLEN: East.
SULLEN: West—far as the Poles asunder.

GEORGE FARQUHAR, *The Beaux' Stratagem*, 1707

A PITY. WE WERE SUCH A GOOD INVENTION

They amputated
Your thighs off my hips.
As far as I'm concerned
They are all surgeons. All of them.

They dismantled us
Each from the other.
As far as I'm concerned
They are all engineers. All of
them.

A pity. We were such a good
And loving invention.
An aeroplane made from a man
and wife.
Wings and everything.
We hovered a little above the
earth.

We even flew a little.

YEHUDA AMICHAI, 1963–8

THE DIVORCE

First it was only an imperceptible quivering of the skin—
'As you please'—where the flesh is darkest.
'Anything wrong?'—Nothing. Milky dreams
of embraces, but the next morning
the other looks different, strangely bony.

Razor-sharp misunderstandings. 'That day in Rome . . .'
I've never said that.—Violent heartbeat,
a kind of hatred, strange.—'That's not the point.'
Repetitions. Glaringly clear the certainty:
From here on, everything is wrong. Odourless and sharp
as a passport photo, this unknown person
with the teacup at the table, with fixed eyes.
It's no use no use no use
—litany in the head, a touch of nausea.
End of reproaches. Slowly the whole room
up to the ceiling fills with guilt.
The plaintive voice is alien, only the shoes
crashing to the floor, the shoes are not.
Next time, in an empty restaurant,
slow-motion, bread crumbs, the subject is money,
laughingly. The dessert tastes of metal.
Two untouchables. Shrill reason.
'Things could be worse.' But at night,
the vindictiveness, the silent battle, anonymously,
like two bony lawyers, two giant crabs
in the water. Then the exhaustion. Slowly
the scab flaking off. A new tobacco store,
a new address. Pariahs, terribly relieved.
Shadows growing paler. This is the file.
This is the key-ring. This is the scar.

HANS MAGNUS ENZENSBERGER, 1985

By this he knew she wept with waking eyes:
That, at his hand's light quiver by her head,
The strange low sobs that shook their common bed
Were called into her with a sharp surprise,
And strangled mute, like little gaping snakes,
Dreadfully venomous to him. She lay
Stone-still, and the long darkness flowed away
With muffled pulses. Then, as midnight makes
Her giant heart of Memory and Tears
Drink the pale drug of silence, and so beat
Sleep's heavy measure, they from head to feet
Were moveless, looking through their dead black years,

By vain regret scrawled over the blank wall.
Like sculptured effigies they might be seen
Upon their marriage-tomb, the sword between;
Each wishing for the sword that severs all.

*

Madam would speak with me. So, now it comes:
The Deluge or else Fire! She's well; she thanks
My husbandship. Our chain on silence clanks.
Time leers between, above his twiddling thumbs.
Am I quite well? Most excellent in health!
The journals, too, I diligently peruse.
Vesuvius is expected to give news:
Niagara is no noisier. By stealth
Our eyes dart scrutinizing snakes. She's glad
I'm happy, says her quivering under-lip.
'And are not you?' 'How can I be?' 'Take ship!
For happiness is somewhere to be had.'
'Nowhere for me!' Her voice is barely heard.
I am not melted, and make no pretence.
With commonplace I freeze her, tongue and sense.
Niagara or Vesuvius is deferred.

It is no vulgar nature I have wived.
Secretive, sensitive, she takes a wound
Deep to her soul, as if the sense had swooned,
And not a thought of vengeance had survived.
No confidences has she: but relief
Must come to one whose suffering is acute.
O have a care of natures that are mute!
They punish you in acts: their steps are brief.
What is she doing? What does she demand
From Providence or me? She is not one
Long to endure this torpidly, and shun
The drugs that crowd about a woman's hand.
At Forfeits during snow we played, and I
Must kiss her. 'Well performed!' I said: then she:
''Tis hardly worth the money, you agree?'
Save her? What for? To act this wedded lie!

*

Thus piteously Love closed what he begat
The union of this ever-diverse pair!
These two were rapid falcons in a snare,
Condemned to do the flitting of the bat.
Lovers beneath the singing sky of May,
They wandered once; clear as the dew on flowers:
But they fed not on the advancing hours:
Their hearts held cravings for the buried day.
Then each applied to each that fatal knife,
Deep questioning, which probes to endless dole.
Ah, what a dusty answer gets the soul
When hot for certainties in this our life!—
In tragic hints here see what evermore
Moves dark as yonder midnight ocean's force,
Thundering like ramping hosts of warrior horse,
To throw that faint thin line upon the shore!

GEORGE MEREDITH, *Modern Love*, 1862

UNIDENTIFIED GUEST: Then no doubt it's all for the best.
With another man, she might have made a mistake
And want to come back to you. If another woman,
She might decide to be forgiving
And gain an advantage. If there's no other woman
And no other man, then the reason may be deeper
And you've ground for hope that she won't come back at all.
If another man, then you'd want to re-marry
To prove to the world that somebody wanted you;
If another woman, you might have to marry her—
You might even imagine that you wanted to marry her.
EDWARD: But I want my wife back.

*

UNIDENTIFIED GUEST: I will say then, you experience some relief
Of which you're not aware. It will come to you slowly:
When you wake in the morning, when you go to bed at night,
That you are beginning to enjoy your independence;
Finding your life becoming cosier and cosier
Without the consistent critic, the patient misunderstander
Arranging life a little better than you like it,

Preferring not quite the same friends as yourself,
Or making your friends like her better than you;
And, turning the past over and over,
You'll wonder only that you endured it for so long.
And perhaps at times you will feel a little jealous
That she saw it first, and had the courage to break it—
Thus giving herself a permanent advantage.

EDWARD: It might turn out so, yet . . .

UNIDENTIFIED GUEST: Are you going to say, you love her?

EDWARD: Why, I thought we took each other for granted.
I never thought I should be any happier
With another person. Why speak of love?
We were used to each other. So her going away
At a moment's notice, without explanation,
Only a note to say that she had gone
And was not coming back—well, I can't understand it.
Nobody likes to be left with a mystery:
It's so . . . unfinished.

UNIDENTIFIED GUEST: Yes, it's unfinished;
And nobody likes to be left with a mystery.
But there's more to it than that. There's a loss of personality;
Or rather, you've lost touch with the person
You thought you were. You no longer feel quite human.
You're suddenly reduced to the status of an object—
A living object, but no longer a person.
It's always happening, because one is an object
As well as a person. But we forget about it
As quickly as we can. When you've dressed for a party
And are going downstairs, with everything about you
Arranged to support you in the role you have chosen,
Then sometimes, when you come to the bottom step
There is one step more than your feet expected
And you come down with a jolt. Just for a moment
You have the experience of being an object
At the mercy of a malevolent staircase.

EDWARD: To what does this lead?

UNIDENTIFIED GUEST: To finding out
What you really are. What you really feel.
What you really are among other people.
Most of the time we take ourselves for granted,

As we have to, and live on a little knowledge
About ourselves as we were. Who are you now?
You don't know any more than I do,
But rather less. You are nothing but a set
Of obsolete responses. The one thing to do
Is to do nothing. Wait.

*

EDWARD: I see now why I wanted my wife to come back.
It was because of what she had made me into.
We had not been alone again for fifteen minutes
Before I felt, and still more acutely—
Indeed, acutely, perhaps, for the first time,
The whole oppression, the unreality
Of the role she had always imposed upon me
With the obstinate, unconscious, sub-human strength
That some women have. Without her, it was vacancy.
When I thought she had left me, I began to dissolve,
To cease to exist. That was what she had done to me!
I cannot live with her—that is now intolerable;
I cannot live without her, for she has made me incapable
Of having any existence of my own.
That is what she has done to me in five years together!
She has made the world a place I cannot live in
Except on her terms. I must be alone,
But not in the same world.

*

I once experienced the extreme of physical pain,
And now I know there is suffering worse than that.
It is surprising, if one had time to be surprised:
I am not afraid of the death of the body,
But this death is terrifying. The death of the spirit—
Can you understand what I suffer?

T. S. ELIOT, *The Cocktail Party*, 1950

MARRIAGES

How dumb before the poleaxe they sink down,
Jostled along the slaughterer's narrow way
To where he stands and smites them one by one.

And now my feet tread that congealing floor,
Encumbered with their offal and their dung,
As each is lugged away to fetch its price.

Carnivorous gourmets, fanciers of flesh,
The connoisseurs of butcher-meat—even these
Must blanch a little at such rituals:

The carcasses of marriages of friends,
Dismemberment and rending, breaking up
Limbs, sinews, joints, then plucking out the heart.

Let no man put asunder . . . Hanging there
On glistening hooks, husbands and wives are trussed,
Silent, and broken, and made separate

By hungers never known or understood,
By agencies beyond the powers they had,
By actions pumping fear into my blood.

ANTHONY THWAITE, 1977

Divorce and suicide have many characteristics in common and one crucial difference: although both are devastatingly public admissions of failure, divorce, unlike suicide, has to be lived through.

A. ALVAREZ, *Life After Marriage*, 1982

10

Growing Old Together

'WILL you still need me, will you still feed me, when I'm sixty-four?' the Beatles sang when they were in their twenties, and 64 seemed a great age. To be allowed to grow old together, to continue to need and to feed one another, 'and the years forgot', must be every couple's hope. Agatha Christie was blessed in her marriage to the archaeologist, Professor Mallowan: 'An archaeologist', she declared, 'is the best husband any woman can have—the older she gets, the more interested he is in her.'

But by no means every aged husband is a Darby 'with Joan by his side', and 'never happy asunder'. Watching and accepting a partner's physical and mental ageing can be intolerably painful, as Thomas Hardy's disturbing poem foreshadows.

Old quarrels, disappointments, and disillusionments sometimes break through with volcanic force as inhibitions are abandoned, defences crumble, and idiosyncracies and egotism grow stronger. The turbulence of the Tolstoys' marriage in later years is a heart-breaking example.

But there are also many couples who enjoy a fruitful old age together. A friend visiting Blake and his wife in their later years, said they were 'as fond of each other, as if their honey moon were still shining . . . they seem animated by one soul, and that a soul of indefatigable industry and benevolence.'

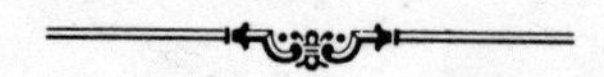

Therefore mercifully ordain that we may become aged together.

The Book of Tobit

A MARRIAGE RING

The ring so worn as you behold,
So thin, so pale, is yet of gold:
The passion such it was to prove;
Worn with life's cares, love yet was love.

GEORGE CRABBE (1754–1832)

TO HIS WIFE

Love, let us live as we have lived, nor lose
  The little names that were the first night's grace,
And never come the day that sees us old,
  I still your lad, and you my little lass.

Let me be older than old Nestor's years,
  And you the Sibyl, if we heed it not.
What should we know, we two, of ripe old age?
  We'll have its richness, and the years forgot.

AUSONIUS, 4th century AD

JOHN ANDERSON

John Anderson my jo, John,
When we were first acquent,
Your locks were like the raven,
Your bonnie brow was brent;
But now your brow is beld, John,
Your locks are like the snow;
But blessings on your frosty pow,
John Anderson my jo.

John Anderson my jo, John,
We clamb the hill thegither,
And mony a canty day, John,
We've had wi' ane anither:
Now we maun totter down, John,
But hand in hand we'll go,
And sleep thegither at the foot,
John Anderson my jo.

ROBERT BURNS, 1790

TO MY WIFE

Dearest, I am getting seedy,
Fat and fussy, kind of greedy.
If your love is on the wane
I can't reasonably complain.
Yet, since legally you're mine,
Try to be my Valentine.

Year by year, to my delight,
You have broiled my chop at night,
Made the toast, and filled my cup.
Oh my darling, keep it up.
Warm my slippers ere we dine.
Damn it, be my Valentine.

CLARENCE DAY, 1936

FIRST VOICE: Mr and Mrs Cherry Owen, in their Donkey Street room that is bedroom, parlour, kitchen, and scullery, sit down to last night's supper of onions boiled in their overcoats and broth of spuds and baconrind and leeks and bones.

MRS CHERRY OWEN: See that smudge on the wall by the picture of Auntie Blossom? That's where you threw the sago.

[*Cherry Owen laughs with delight*

MRS CHERRY OWEN: You only missed me by a inch.

CHERRY OWEN: I always miss Auntie Blossom too.

MRS CHERRY OWEN: Remember last night? In you reeled, my boy, as drunk as a deacon with a big wet bucket and a fish-frail full of stout and you looked at me and you said, 'God has come home!' you said, and then over the bucket you went, sprawling and bawling, and the floor was all flagons and eels.

CHERRY OWEN: Was I wounded?

MRS CHERRY OWEN: And then you took off your trousers and you said, 'Does anybody want a fight!' Oh, you old baboon.

CHERRY OWEN: Give me a kiss.

MRS CHERRY OWEN: And then you sang 'Bread of Heaven', tenor and bass.

CHERRY OWEN: I *always* sing 'Bread of Heaven'.

MRS CHERRY OWEN: And then you did a little dance on the table.

CHERRY OWEN: I did?

MRS CHERRY OWEN: Drop dead!

CHERRY OWEN: And then what did I do?

MRS CHERRY OWEN: Then you cried like a baby and said you were a poor drunk orphan with nowhere to go but the grave.

CHERRY OWEN: And what did I do next, my dear?

MRS CHERRY OWEN: Then you danced on the table all over again and said you were King Solomon Owen and I was your Mrs Sheba.

CHERRY OWEN (*Softly*): And then?

MRS CHERRY OWEN: And then I got you into bed and you snored all night like a brewery.

[*Mr and Mrs Cherry Owen laugh delightedly together*

DYLAN THOMAS, *Under Milk Wood*, 1954

LOVE'S ANNIVERSARIES

(FOR JOYCE)

It was the generosity of delight
that first we learned in a sparsely-furnished flat
clothed in our lovers' nakedness. By night
we timidly entered what we marvelled at,

ranging the flesh's compass. But by day
we fell together, fierce with awkwardness
that window-light and scattered clothing lay
impassive round such urgent happiness.

Now, children, years and many rooms away,
and tired with experience, we climb the stairs
to our well-furnished room; undress, and say
familiar words for love; and from the cares

that back us, turn together and once more seek
the warmth of wonder each to the other meant
so strong ago, and with known bodies speak
the unutterable language of content.

MAURICE LINDSAY, 1964

How it is I know not; but there is no place like a bed for confidential disclosures between friends. Man and wife, they say, there open the very bottom of their souls to each other; and some old couples often lie and chat over old times till nearly morning.

HERMAN MELVILLE, *Moby Dick*, 1851

At the sea-end of town, Mr and Mrs Floyd, the cocklers, are sleeping as quiet as death, side by wrinkled side, toothless, salt and brown, like two old kippers in a box.

DYLAN THOMAS, *Under Milk Wood*, 1954

But if you survived melancholia and rotting lungs it was possible to live long in this valley. Joseph and Hannah Brown, for instance, appeared to be indestructible. For as long as I could remember they had lived together in the same house by the common. They had lived there, it was said, for fifty years; which seemed to me for ever. They had raised a large family and sent them into the world, and had continued to live on alone, with nothing left of their noisy brood save some dog-eared letters and photographs.

The old couple were as absorbed in themselves as lovers, content and self-contained; they never left the village or each other's company, they lived as snug as two podded chestnuts. By day blue smoke curled up from their chimney, at night the red windows glowed; the cottage, when we passed it, said 'Here live the Browns', as though that were part of nature.

Though white and withered, they were active enough, but they ordered their lives without haste. The old woman cooked, and threw grain to the chickens, and hung out her washing on bushes; the old man fetched wood and chopped it with a billhook, did a bit of gardening now and then, or just sat on a seat outside his door and gazed at the valley, or slept. When summer came they bottled fruit, and when winter came they ate it. They did nothing more than was necessary to live, but did it fondly, with skill—then sat together in their clock-ticking kitchen enjoying their half-century of silence. Whoever called to see them was welcomed gravely, be it man or beast or child; and to me they resembled two tawny insects, slow but deft in their movements; a little foraging, some frugal feeding, then any amount of stillness. They spoke to each other without raised voices, in short chirrups as brief as bird-song, and when they moved about in their tiny kitchen they did so smoothly and blind, gliding on worn, familiar rails, never bumping or obstructing each other. They were fond, pink-faced, and alike as cherries, having taken and merged, through their years together, each other's looks and accents.

It seemed that the old Browns belonged for ever, and that the miracle of their survival was made commonplace by the durability of their love—if one should call it love, such a balance. Then suddenly, within the space of two days, feebleness took them both. It was as though two machines, wound up and synchronized, had run down at exactly the same time. Their interdependence was so legendary we didn't notice their plight at first. But after a week, not having been seen about, some neighbours thought it best to call. They found old Hannah on the kitchen floor feeding her man with a spoon. He was lying in a corner half-covered with matting, and they were both too weak to stand. She had chopped up a plate of peelings, she said, as she hadn't been able to manage the fire. But they were all right really, just a touch of the damp; they'd do, and it didn't matter.

Well, the Authorities were told; the Visiting Spinsters got busy; and it was decided they would have to be moved. They were too frail to help each other now, and their children were too scattered, too busy. There was but one thing to be done; it was for the best; they would have to be moved to the Workhouse.

The old couple were shocked and terrified, and lay clutching each other's hands. 'The Workhouse'—always a word of shame, grey shadow falling on the close of life, most feared by the old (even when called The Infirmary); abhorred more than debt, or prison, or beggary, or even the stain of madness.

Hannah and Joseph thanked the Visiting Spinsters but pleaded to be left at home, to be left as they wanted, to cause no trouble, just simply to stay together. The Workhouse could not give them the mercy they needed, but could only divide them in charity. Much better to hide, or die in a ditch, or to starve in one's familiar kitchen, watched by the objects one's life had gathered—the scrubbed empty table, the plates and saucepans, the cold grate, the white stopped clock. . . .

'You'll be well looked after,' the Spinsters said, 'and you'll see each other twice a week.' The bright busy voices cajoled with authority and the old couple were not trained to defy them. So that same afternoon, white and speechless, they were taken away to the Workhouse. Hannah Brown was put to bed in the Women's Wing, and Joseph lay in the Men's. It was the first time, in all their fifty years, that they had ever been separated. They did not see each other again, for in a week they both were dead.

I was haunted by their end as by no other, and by the kind, killing Authority that arranged it. Divided, their life went out of them, so they ceased as by mutual agreement. Their cottage stood empty on the edge of the common, its front door locked and soundless. Its stones grew rapidly cold and repellent with its life so suddenly withdrawn. In a year it fell down, first the roof, then the walls, and lay scattered in a tangle of briars. Its decay was so violent and overwhelming, it was as though the old couple had wrecked it themselves.

Soon all that remained of Joe and Hannah Brown, and of their long close life together, were some grass-grown stumps, a garden gone wild, some rusty pots, and a dog-rose.

LAURIE LEE, *Cider with Rosie*, 1959

'IN THE NIGHT SHE CAME'

I told her when I left one day
That whatsoever weight of care
Might strain our love, Time's mere assault
    Would work no changes there.
And in the night she came to me,
    Toothless, and wan, and old,
With leaden concaves round her eyes,
    And wrinkles manifold.

I tremblingly exclaimed to her,
'O wherefore do you ghost me thus!
I have said that dull defacing Time
Will bring no dreads to us.'
'And is that true of *you*?' she cried
In voice of troubled tune.
I faltered: 'Well . . . I did not think
You would test me quite so soon!'

She vanished with a curious smile,
Which told me, plainlier than by word,
That my staunch pledge could scarce beguile
The fear she had averred.
Her doubts then wrought their shape in me,
And when next day I paid
My due caress, we seemed to be
Divided by some shade.

THOMAS HARDY, 1909

At the present time, in the dark little parlour certain feet below the level of the street—a grim, hard, uncouth parlour, only ornamented with the coarsest of baize table-covers, and the hardest of sheet-iron tea-trays, and offering in its decorative character no bad allegorical representation of Grandfather Smallweed's mind—seated in two black horse-hair porter's chairs, one in each side of the fireplace, the superannuated Mr and Mrs Smallweed wile away the rosy hours. On the stove are a couple of trivets for the pots and kettles which it is Grandfather Smallweed's usual occupation to watch, and projecting from the chimney-piece between them is a sort of brass gallows for roasting, which he also superintends when it is in action. Under the venerable Mr Smallweed's seat, and guarded by his spindle legs, is a drawer in his chair, reported to contain property to a fabulous amount. Beside him is a spare cushion, with which he is always provided, in order that he may have something to throw at the venerable partner of his respected age whenever she makes an allusion to money—a subject on which he is particularly sensitive.

'And where's Bart?' Grandfather Smallweed inquires of Judy, Bart's twin-sister.

'He an't come in yet,' says Judy.

'It's his tea-time, isn't it?'

'No.'

'How much do you mean to say it wants then?'

'Ten minutes.'

'Hey?'

'Ten minutes.'—(Loud on the part of Judy.)

'Ho!' says Grandfather Smallweed. 'Ten minutes.'

Grandmother Smallweed, who has been mumbling and shaking her head at the trivets, hearing figures mentioned, connects them with money, and screeches, like a horrible old parrot without any plumage, 'Ten ten-pound notes!'

Grandfather Smallweed immediately throws the cushion at her.

'Drat you, be quiet!' says the good old man.

The effect of this act of jaculation is twofold. It not only doubles up Mrs Smallweed's head against the side of her porter's chair, and causes her to present, when extricated by her granddaughter, a highly unbecoming state of cap, but the necessary exertion recoils on Mr Smallweed himself, whom it throws back into *his* porter's chair, like a broken puppet. The excellent old gentleman being, at these times, a mere clothes-bag with a black skull-cap on the top of it, does not present a very animated appearance until he has undergone the two operations at the hands of his granddaughter, of being shaken up like a great bottle, and poked and punched like a great bolster. Some indication of a neck being developed in him by these means, he and the sharer's of his life's evening again sit fronting one another in their two porter's chairs, like a couple of sentinels long forgotten on their post by the Black Serjeant, Death.

CHARLES DICKENS, *Bleak House*, 1852–3

For forty-seven years they had been married. How deep back the stubborn gnarled roots of the quarrel reach, no one could say—but only now, when tending to the needs of others no longer shackled them together, the roots swelled up visible, split the earth between them, and the tearing shook even to the children, long since grown.

TILLIE OLSEN, *Tell Me a Riddle*, 1962

AT THE DRAPERS

'I stood at the back of the shop, my dear,
  But you did not perceive me.
Well, when they deliver what you were shown
  I shall know nothing of it, believe me!'

And he coughed and coughed as she paled and said,
  'O, I didn't see you come in there—
Why couldn't you speak?'—'Well, I didn't. I left
  That you should not notice I'd been there.

You were viewing some lovely things. "*Soon required
  For a widow, of latest fashion*";
And I knew 'twould upset you to meet the man
  Who had to be cold and ashen

And screwed in a box before they could dress you
  "*In the last new note in mourning*,"
As they defined it. So, not to distress you,
  I left you to your adorning.'

THOMAS HARDY, 1911

For close on half a century we two had been enemies, and now, on this heavy afternoon, the enemies had suddenly become aware of the bond created, in spite of the long-drawn-out struggle, by a shared old age. We might seem to hate one another, but, for all that, we had reached the same point in the road. There was nothing now beyond that promontory on which we stood awaiting death. . . .

Husbands and wives of long standing never hate one another as much as they think they do.

FRANÇOIS MAURIAC, *The Knot of Vipers*, 1932

21 February 1895: I am passing through yet another painful period. I don't even want to write about it, it is so terrible, so difficult and so clear to me now that my life is going into a decline. I have no desire to live and thoughts of suicide pursue me ever more relentlessly. Save me Lord from such a sin! Today I again tried to leave home; I think I

must be ill, I cannot control myself. All my sufferings are as nothing compared to the one great grief in my soul—Lyovochka's indifference to me and the children. Surely there are some happy old couples, who have loved each other as passionately as we have for the past 33 years, and have now developed a relationship based on friendship and affection? As for us—I keep having stupid outbursts of sentimental passion for him. When I was ill he brought me 2 wonderful apples, and I planted the pips in memory of this rare display of tenderness towards me. Will I ever see those pips grow into trees, I wonder.

Yes, I was going to relate the whole dreadful *episode* between us. It is all my fault, of course, yet how did I get dragged into it in the first place? Just so long as the children don't blame me, for no one will ever be able to understand our marital relations. If, in spite of my apparent happiness, I want to end my life, and have attempted so many times to do so, surely there must be some reason for it? If only people knew how painful it is—these endless outbursts, these attempts at love, which is painfully wearing out, for it never receives anything but physical gratification; and even more painful is the realisation in the *last* days of our life together, that there are no mutual feelings between us, that for the whole of my life I have single-mindedly and unwaveringly loved a man who was utterly selfish and returned all my feelings with a withering and pitiless scorn.

SOFIA TOLSTOY, *Diaries*

AFTER THE GOLDEN WEDDING

(Three Soliloquies)

I THE HUSBAND'S

She's not a faultless woman; no!
  She's not an angel in disguise:
She has her rivals here below:
  She's not an unexampled prize:

She does not always see the point
  Of little jests her husband makes:
And, when the world is out of joint,
  She makes a hundred small mistakes:

She's not a miracle of tact:
  Her temper's not the best I know:
She's got her little faults in fact,
  Although I never tell her so.

But this, my wife, is why I hold you
  As good a wife as ever stepped,
And why I meant it when I told you
  How cordially our feast I kept:

You've lived with me these fifty years,
  And all the time you loved me dearly:
I may have given you cause for tears:
  I may have acted rather queerly.

I ceased to love you long ago:
  I loved another for a season:
As time went on I came to know
  Your worth, my wife: and saw the reason

Why such a wife as you have been
  Is more than worth the world beside;
You loved me all the time, my Queen;
  You couldn't help it if you tried.

You loved me as I once loved you,
  As each loved each beside the altar:
And whatsoever I might do,
  Your loyal heart could never falter.

And, if you sometimes fail me, sweetest,
  And don't appreciate me, dear,
No matter: such defects are meetest
  For poor humanity, I fear.

And all's forgiven, all's forgot,
  On this our golden wedding day;
For, see! she loves me: does she not?
  So let the world e'en go its way.

I'm old and nearly useless now,
   Each day a greater weakling proves me:
There's compensation anyhow:
   I still possess a wife that loves me.

II THE WIFE'S

Dear worthy husband! good old man!
   Fit hero of a golden marriage:
I'll show towards you, if I can,
   An absolutely wifely carriage.

The months or years which your career
   May still comprise before you perish,
Shall serve to prove that I, my dear,
   Can honour, and obey, and cherish.

Till death us part, as soon he must,
   (And you, my dear, should show the way)
I hope you'll always find me just
   The same as on our wedding day.

I never loved you, dearest: never!
   Let that be clearly understood:
I thought you good, and rather clever,
   And found you really rather good.

And, what was more, I loved another,
   But couldn't get him: well, but, then
You're just as bad, my erring brother,
   You most impeccable of men:—

Except for this: my love was married
   Some weeks before I married you:
While you, my amorous dawdler, tarried
   Till we'd been wed a year or two.

You loved me at our wedding: I
   Loved some one else: and after that
I never cast a loving eye
   On others: you—well, tit for tat!

But after all I made you cheerful:
  Your whims I've humoured: saw the point
Of all your jokes: grew duly tearful,
  When you were sad, yet chose the joint

You like the best of all for dinner,
  And soothed you in your hours of woe:
Although a miserable sinner,
  I *am* a good wife, as wives go.

I bore with you and took your side,
  And kept my temper all the time:
I never flirted; never cried,
  Nor ranked it as a heinous crime,

When you preferred another lady,
  Or used improper words to me,
Or told a story more than shady,
  Or snored and snorted after tea,

Or otherwise gave proofs of being
  A dull and rather vain old man:
I still succeeded in agreeing
  With all you said, (the safest plan),

Yet always strove my point to carry,
  And make you do as I desired:
I'm *glad* my people made me marry!
  They hit on just what I required.

Had love been wanted—well, I couldn't
  Have given what I'd not to give;
Or had a genius asked me! wouldn't
  The man have suffered? now, we live

Among our estimable neighbours
  A decent and decorous life:
I've earned by my protracted labours
  The title of a model wife.

But when beneath the turf you're sleeping,
  And I am sitting here in black,
Engaged, as they'll suppose, in weeping,
  I shall not wish to have you back.

III THE VICAR'S

A good old couple! kind and wise!
  And oh! what love for one another!
They've won, those two, life's highest prize,
  Oh! let us copy them, my brother.

J. K. STEPHEN, 1891

Mr Bulstrode, prosperous and sanctimonious, has been publicly exposed for a long-buried swindle, and his wife has just heard of his disgrace:

She locked herself in her room. She needed time to get used to her maimed consciousness, her poor lopped life, before she could walk steadily to the place allotted her. A new searching light had fallen on her husband's character, and she could not judge him leniently: the twenty years in which she had believed in him and venerated him by virtue of his concealments came back with particulars that made them seem an odious deceit. He had married her with that bad past life hidden behind him and she had no faith left to protest his innocence of the worst that was imputed to him. Her honest ostentatious nature made the sharing of a merited dishonour as bitter as it could be to any mortal.

But this imperfectly-taught woman, whose phrases and habits were an odd patchwork, had a loyal spirit within her. The man whose prosperity she had shared through nearly half a life, and who had unvaryingly cherished her—now that punishment had befallen him it was not possible to her in any sense to forsake him. There is a forsaking which still sits at the same board and lies on the same couch with the forsaken soul, withering it the more by unloving proximity. She knew, when she locked her door, that she should unlock it ready to go down to her unhappy husband and espouse his sorrow, and say of his guilt, I will mourn and not reproach. But she needed time to gather up her strength; she needed to sob out her farewell to all the gladness and pride of her life. When she had resolved to go down, she prepared herself by some little acts which might seem mere folly to a hard onlooker; they were her way of expressing to all spectators visible or invisible that she had begun a new life in which she embraced humiliation. She took off all her ornaments and put on a plain black gown, and instead of wearing her much-adorned cap and large bows of hair, she brushed her hair down and put on a plain bonnet-cap, which made her look suddenly like an early Methodist.

Bulstrode, who knew that his wife had been out and had come in saying that she was not well, had spent the time in an agitation equal to hers. He had looked forward to her learning the truth from others, and had acquiesced in that probability, as something easier to him than any confession. But now that he imagined the moment of her knowledge come, he awaited the result in anguish. His daughters had been obliged to consent to leave him, and though he had allowed some food to be brought to him, he had not touched it. He felt himself perishing slowly in unpitied misery. Perhaps he should never see his wife's face with affection in it again. And if he turned to God there seemed to be no answer but the pressure of retribution.

It was eight o'clock in the evening before the door opened and his wife entered. He dared not look up at her. He sat with his eyes bent down, and as she went towards him she thought he looked smaller—he seemed so withered and shrunken. A movement of new compassion and old tenderness went through her like a great wave, and putting one hand on his which rested on the arm of the chair, and the other on his shoulder, she said solemnly but kindly, 'Look up, Nicholas.'

He raised his eyes with a little start and looked at her half amazed for a moment: her pale face, her changed, mourning dress, the trembling about her mouth, all said, 'I know'; and her hands and eyes rested gently on him. He burst out crying and they cried together, she sitting at his side. They could not yet speak to each other of the shame which she was bearing with him or of the acts which had brought it down on them. His confession was silent, and her promise of faithfulness was silent. Open-minded as she was, she nevertheless shrank from the words which would have expressed their mutual consciousness as she would have shrunk from flakes of fire. She could not say, 'How much is only slander and false suspicion?' And he did not say, 'I am innocent.'

GEORGE ELIOT, *Middlemarch*, 1871–2

WHEN YOU ARE OLD

When you are old and grey and full of sleep,
And nodding by the fire, take down this book,
And slowly read, and dream of the soft look
Your eyes had once, and of their shadows deep;

How many loved your moments of glad grace,
And loved your beauty with love false or true,
But one man loved the pilgrim soul in you,
And loved the sorrows of your changing face;

And bending down beside the glowing bars,
Murmur, a little sadly, how Love fled
And paced upon the mountains overhead
And hid his face amid a crowd of stars.

W. B. YEATS, 1906

GRANDPARENTS

Old and gnarled; wizened faces dreaming of
Past days gone by.
Living in their own world:
Christmasses; presents;
Easter; church service on television;
Telling of experiences, adding bits here and there
To improve the story;
Surprised at the cost of living; not understanding
Why prices go up and up;
Enjoying company, but company not enjoying
Them. Yet happy in their own way.

SHEILA BRAMFIT, from *Children as Poets*, 1972

AN ARUNDEL TOMB

Side by side, their faces blurred,
The earl and countess lie in stone,
Their proper habits vaguely shown

As jointed armour, stiffened pleat,
And that faint hint of the absurd—
The little dogs under their feet.

Such plainness of the pre-baroque
Hardly involves the eye, until
It meets his left-hand gauntlet, still
Clasped empty in the other; and
One sees, with a sharp tender shock,
His hand withdrawn, holding her hand.

They would not think to lie so long.
Such faithfulness in effigy
Was just a detail friends would see:
A sculptor's sweet commissioned grace
Thrown off in helping to prolong
The Latin names around the base.

They would not guess how early in
Their supine stationary voyage
The air would change to soundless damage,
Turn the old tenantry away;
How soon succeeding eyes begin
To look, not read. Rigidly they

Persisted, linked, through lengths and breadths
Of time. Snow fell, undated. Light
Each summer thronged the glass. A bright
Litter of birdcalls strewed the same
Bone-riddled ground. And up the paths
The endless altered people came,

Washing at their identity.
Now, helpless in the hollow of
An unarmorial age, a trough
Of smoke in slow suspended skeins
Above their scrap of history,
Only an attitude remains:

Time has transfigured them into
Untruth. The stone fidelity
They hardly meant has come to be
Their final blazon, and to prove
Our almost-instinct almost true:
What will survive of us is love.

PHILIP LARKIN, 1964

THE ANNIVERSARY

All Kings, and all their favourites,
All glory of honours, beauties, wits,
The sun itself, which makes times, as they pass,
Is elder by a year now than it was
When thou and I first one another saw:
All other things to their destruction draw,
Only our love hath no decay;
This no tomorrow hath, nor yesterday,
Running it never runs from us away,
But truly keeps his first, last, everlasting day.

Two graves must hide thine and my corse;
If one might, death were no divorce.
Alas, as well as other Princes, we
(Who Prince enough in one another be)
Must leave at last in death these eyes and ears,
Oft fed with true oaths, and with sweet salt tears;
But souls where nothing dwells but love
(All other thoughts being inmates) then shall prove
This, or a love increasèd there above,
When bodies to their graves, souls from their graves remove.

And then we shall be throughly blessed;
But we no more than all the rest.
Here upon earth we're Kings, and none but we
Can be such Kings, nor of such subjects be;
Who is so safe as we? where none can do
Treason to us, except one of us two.

True and false fears let us refrain,
Let us love nobly, and live, and add again
Years and years unto years, till we attain
To write threescore: this is the second of our reign.

JOHN DONNE, pub. 1633

IDLENESS

I keep the rustic gate closed
For fear somebody might step
On the green moss. The sun grows
Warmer. You can tell it's Spring.
Once in a while, when the breeze
Shifts, I can hear the sounds of the
Village. My wife is reading
The classics. Now and then she
Asks me the meaning of a word.
I call for wine and my son
Fills my cup till it runs over.
I have only a little
Garden, but it is planted
With yellow and purple plums.

LU YU, *c.*1200

BAUCIS AND PHILEMON

The gods' power is boundless; whatever they decree comes true. The proof is on a hillside in Phrygia, where a linden and an oak-tree grow entwined, beside a wall. Not far away is a marsh, the home of coots, ducks, and other water-birds. Once, it was prosperous farmland. But then two gods, Jupiter and Mercury, passed that way, disguised as human travellers. They asked for shelter at a thousand doors, and a thousand householders took them for tramps and set the dogs on them. At last, on a hillside, they came to a tiny, reed-thatched cottage, the home of Baucis and her aged husband Philemon. Baucis and Philemon had married when they were little more than children, and had lived in that lonely place for sixty years, content in each other's company. In that household there were no servants, no employers: each of the two old people gave orders, and each obeyed.

When the gods knocked on the cottage door, still disguised as tramps, Philemon shook their hands and welcomed them inside as if they were long-lost relatives. They bent their heads to pass through the low doorway, and Philemon spread a rug on a bench and invited them to sit and rest after their journey. Baucis stirred the ashes of yesterday's fire, blew them into life and fed the flames with dry reeds and bark. When the fire blazed she piled it with twigs and branches, and put water in a copper pot to boil. Philemon meanwhile fetched a cabbage from the garden, and cut a slice of smoked pork from a leg hanging, soot-blackened, from the ceiling.

While the old couple worked, they chattered as if they'd seen no visitors for years. They filled a beech-wood bowl with water for their guests to wash their weary feet. They plumped the reed mattress on their own bed, and laid over it an embroidered cloth treasured since their marriage, brought out only on solemn occasions. The gods reclined at ease, and Baucis set in front of them a rickety table with a piece of tile under one leg to balance it. She wiped the table-top with a sprig of mint, and set out black and green olives, plum chutney, radishes, endive, cheese, and duck-eggs baked in the ashes. There were beechwood cups, and a clay jar of wine. That was the first course; the second was bacon stew, and the third was a basket of nuts, figs, plums and grapes, with a honeycomb on a wooden dish.

Smiling, the old people watched their guests enjoy the meal. Then, suddenly, they realized that each time the wine-jar was emptied it welled full again of its own accord. Recognizing their visitors for immortal gods, they fell on their knees and begged pardon for the plainness of the meal. Then they remembered their goose, the guardian of the house. They ran to catch him and sacrifice him for the gods, their guests. But they were old and slow; the goose flapped easily out of reach, and flew for sanctuary to the gods themselves. 'Enough,' said Jupiter. 'We are gods, and though this whole area will be destroyed, your kindness has moved us to spare your lives. Leave the cottage; climb the hill with us.'

The old people struggled up the hill behind the gods, leaning on their sticks. A bowshot from the top they turned to look back. Behind them, the land and farms had disappeared. There was nothing to be seen but marsh, and their cottage alone on the hillside. Even as they watched, weeping for their belongings, the cottage walls were transformed into stone pillars, the floor turned to marble and the thatch became yellow gold. 'Old man, old woman,' said Jupiter, 'tell us what you desire most in all the world.'

Baucis and Philemon murmured together for a moment, and then Philemon said, 'Let us be your priests and look after this your temple. And since we have lived all our lives peacefully together, let me not live to bury Baucis, nor Baucis live to bury me. Let the same moment end both our lives.'

Their prayer was answered. They lived as priest and priestess of the temple for many more years, honouring the gods. Then one day, as they stood quietly together by the temple steps, talking of old times, Baucis suddenly saw tendrils sprouting from Philemon's shoulders, and at the same moment he noticed that the skin of her arms was growing brown and hard, like bark. They tried to move their feet, and found that they were taking root in the ground they stood on. Baucis was becoming a linden and Philemon an oak—and they had just time to kiss one another for the last time and say 'Farewell, my dear', before the transformation was complete.

OVID, *Metamorphoses* VIII, AD 8
adapted by Kenneth McLeish, 1987

II

Till Death us do Part

THE death of a spouse is an event which must sooner or later occur in every lasting marriage; bereavement, as C. S. Lewis says, follows marriage as marriage follows courtship.

Each survivor will experience bereavement differently. But there is a process of grief, a 'long, winding valley', which all mourners must go through. C. S. Lewis, in his diary, has charted the stages of this journey, from the agony of despair, the refusal to accept the reality of the death (which sometimes takes the form of trying to get in touch with the dead), through exhaustion, to an unexpected lifting of the darkness as the memory of the loved person as he or she was in life returns.

Desmond MacCarthy wrote to a friend:

> It seems likely that one cannot remember the dead truly—the essence of their being—until one has recovered in some measure the enjoyment of life. Their death was not the characteristic thing about them. It was their response to life, and it is by living oneself again that one meets them as they were and keeps in closest touch with them.

REMEMBER ME

Remember me when I am gone away,
Gone far away into the silent land;
When you can no more hold me by the hand,
Nor I half turn to go, yet turning stay.
Remember me when no more day by day
You tell me of our future that you plann'd:
Only remember me; you understand
It will be late to counsel then or pray.
Yet if you should forget me for a while
And afterwards remember, do not grieve:
For if the darkness and corruption leave
A vestige of the thoughts that once I had,
Better by far you should forget and smile
Than that you should remember and be sad.

CHRISTINA ROSSETTI, 1862

Do not grieve for me too much, I am a spirit confident of my rights. Death is only an incident, & not the most important wh happens to us in this state of being. On the whole, especially since I met you my darling one I have been happy, & you have taught me how noble a womans heart can be. If there is anywhere else I shall be on the look out for you. Meanwhile look forward, feel free, rejoice in life, cherish the children, guard my memory. God bless you.

Good bye

W

WINSTON CHURCHILL, *from a letter to his wife on the eve of being sent to the Dardanelles, 17 July 1915.*

GOODBYE

So we must say Goodbye, my darling,
And go, as lovers go, for ever;
Tonight remains, to pack and fix on labels,
And make an end of lying down together.

I put a final shilling in the gas,
And watch you slip your dress below your knees
And lie so still I hear your rustling comb
Modulate the autumn in the trees.

And all the countless things I shall remember
Lay mummy-cloths of silence round my head;
I fill the carafe with a drink of water;
You say: 'We paid a guinea for this bed'.

And then, 'We'll leave some gas, a little warmth
For the next resident, and these dry flowers,'
And turn your face away, afraid to speak
The big word, that Eternity is ours.

Your kisses close my eyes, and yet you stare
As though God struck a child with nameless fears;
Perhaps the water glitters and discloses
Time's chalice, and its limpid useless tears.

Everything we renounce except ourselves.
Selfishness is the last of all to go.
Our sighs are exhalations of the earth,
Our footprints leave a track across the snow.

We made the universe to be our home,
Our nostrils took the wind to be our breath;
Our hearts are massive towers of delight,
We stride across the seven seas of death.

Yet when all's done, you'll keep the emerald
I placed upon your finger in the street;
And I will keep the patches that you sewed
On my old battledress tonight, my sweet.

ALUN LEWIS, 1945

FRANCE

A dozen sparrows scuttled on the frost.
We watched them play. We stood at the window,
And, if you saw us, then you saw a ghost
In duplicate. I tied her night-gown's bow.
She watched and recognized the passers-by.
Had they looked up, they'd know that she was ill—
'Please, do not draw the curtains when I die'—
From all the flowers on the windowsill.

'It's such a shame', she said. 'Too ill, too quick.
I would have liked us to have gone away.'
We closed our eyes together, dreaming France,
Its meadows, rivers, woods and *jouissance*.
I counted summers, our love's arithmetic.
'Some other day, my love. Some other day.'

DOUGLAS DUNN, 1985

BEFORE THE BIRTH OF ONE OF HER CHILDREN

All things within this fading world hath end,
Adversity doth still our joys attend;
No ties so strong, no friends so dear and sweet,
But with death's parting blow is sure to meet.
The sentence past is most irrevocable,
A common thing, yet oh, inevitable.
How soon, my Dear, death may my steps attend,
How soon't may be thy lot to lose thy friend,
We both are ignorant, yet love bids me
These farewell lines to recommend to thee,

That when that knot's untied that made us one,
I may seem thine, who in effect am none.
And if I see not half my days that's due,
What nature would, God grant to yours and you;
The many faults that well you know I have
Let be interred in my oblivious grave;
If any worth or virtue were in me,
Let that live freshly in thy memory
And when thou feel'st no grief, as I no harms,
Yet love thy dead, who long lay in thine arms.
And when thy loss shall be repaid with gains
Look to my little babes, my dear remains.
And if thou love thyself, or loved'st me,
These O protect from step-dame's injury.
And if chance to thine eyes shall bring this verse,
With some sad sighs honour my absent hearse;
And kiss this paper for thy love's dear sake,
Who with salt tears this last farewell did take.

ANNE BRADSTREET (1612–72)

When he did talk he often rambled and said the strange things which frightened her so: that fine clear mind in whose *light* she had lived become helplessly confused and darkened. Perhaps he was silent because he feared to affright her. Or perhaps abhorred this final loss of face before his wife, this unspeakable defeat at the hands of fate. Or would not feed a love which was so soon to be transmuted, he to sleep, she to mourn.

They had always been very close to each other, united by indistinguishably close bonds of love and intelligence. They had never ceased passionately to crave each other's company. They had never seriously quarrelled, never been parted, never doubted each other's complete honesty. A style of directness and truthfulness composed the particular gaiety of their lives. Their love had grown, nourished daily by the liveliness of their shared thoughts. They had grown together in mind and body and soul as it is sometimes blessedly given to two people to do. They could not be in the same room without touching each other. They constantly uttered even their most trivial thoughts. Their converse passed through wit. Jest and reflection had been the language of their love. I shall die without him, thought

Gertrude, not suicide, but I shall just have no more life. I shall be a dead person walking about.

IRIS MURDOCH, *Nuns and Soldiers*, 1980

UPON THE DEATH OF SIR ALBERTUS MORTON'S WIFE

He first deceased; she for a little tried
To live without him, liked it not, and died.

SIR HENRY WOTTON, 1651

Our mother lay in the large front room—Moor in the little room behind and next to it. And they who were so used to each other, whose lives had come to form part of each other, could not be in the same room any longer. Never shall I forget the morning when he felt strong enough to go into mother's room. When they were together they were young again, she a loving girl and he a loving youth, on the threshold of life, not an old man devastated by illness and an old woman parting from each other for life. . . . She remained fully conscious almost to the last moment and when she could no longer speak she pressed our hands and tried to smile . . . but the last word she spoke to Papa was 'good'.

from a letter from Tussy, Karl ('Moor') Marx's daughter, describing her mother's death in 1881

When I got to the hospital the Sister said: 'I am very sorry, I think you are just in time', and began walking at a tremendous speed along a corridor until we came to a small room in which was a very high bed, and nothing else. Elma was lying on the bed inside a plastic tent, living her last few minutes in her own private atmosphere; she was looking with great concentration at nothing, and exactly at the moment when I reached the bedside she sighed very deeply and died. The nurse said: 'What a pity; I am very sorry; she was very young, wasn't she; you just can't tell; now I think you had better go to the waiting-room.' So I went to the waiting-room and that was the end of it, the end of Elma, the end of my short and little marriage, the end of my short and little life. After an hour or so they let me back in again, and I put my flowers somewhere on the place where she lay. The nurse said: 'I think the raid-warning's gone; I am very sorry this happened; we shall look after everything; now perhaps you had better go and leave it to us.'

So I went away and left it to them; I was twenty-eight, and old enough to understand that hospitals were busy places.

*

She was both kind and tranquil, she was beautiful and she was generous; she was as vulnerable as I but more composed, and nobody else did or could have done what I had supposed impossible: she took me over the barrier between the past and the present, and opened all the closed doors. When we were married it was like entering a theatre already in the second act; we united our children—my baby daughter, her baby son—and for the first time for three despairing years there seemed for me some point in establishing a root in life.

JAMES CAMERON, *Point of Departure*, 1967

When she felt the sleeping pill beginning to work she clicked off the lamp and lay on her side away from Tusker's light and shut her eyes. Now my own, my love, she thought. Now my own. Now.

And slept.

Woke, shivering. Her watch showed 3.30. She should not have gone to bed so early. The pill had worked, but worn off. She got out of bed, put on her gown and slippers and was shattered by recollection. Tusker's empty bed. Going past it to the living-room she switched his lamp off. The living-room was unlit except by the light coming from the bedroom. She trod gently. Curled up near the almost dead fire were two shapes in blankets. Minnie and Ibrahim, one on each side of the fireplace. Going gently past them she caught her breath because there was a third shape, huddled with its back to the wall.

Joseph.

The three of them.

No, she told herself. It is very moving, but I mustn't cry. If I cry I may not be able to stop and that will never do, will it? There is such a lot to think about and attend to. She stood in the dark kitchen until her eyes were used to it and she could make out the shape of the brandy bottle and a glass, probably a dirty one, but that didn't matter. She moved cautiously back through the living-room with bottle and glass, anxious not to disturb the sleeping watch.

Back in the bedroom she poured a stiff measure of brandy. To drink it she sat on Tusker's cold bed and stared through the net of her own to the lamp on the other side of it, and then remembered what it was that she had to do tomorrow as well as go to Tusker's funeral. She gulped some of the brandy. . . . She went into the bathroom to run some cold tap-water into it. She ran too much. Now it was too weak. She went back to the bottle and topped the drink up and then back into the bathroom because she suddenly felt sick, and her bowels were stirring. She stood by the basin, waiting to be sick. She was unable to be sick, but her bowels were still moving, so she went over to the thrones and sat on hers, with the brandy glass on the floor within reach. She heard herself moaning quietly and at once stifled the sound with her hand over her mouth. She did not want to wake the servants. She had forgotten to put a towel over the shutters. . . .

Now that she was alone she would have to have the catch put back on the shutters. Tusker had had it taken off for the same reason that he had insisted on two loos. In India, he had said, you could never tell when you'd get taken short. And who could tell if you both might not get taken short at the same moment? If they could only have one bathroom, they could at least have two loos in it and no catch on the door. Actually it had only happened once, the time they'd both eaten something that disagreed with them. She'd always sworn she'd never undergo the indignity of sitting on her loo while Tusker was sitting on his. But, this once, she'd been driven to it, and half way through the performance Tusker had begun to laugh and after a while she had begun to laugh too, so there they had been, enthroned, laughing like drains.

She began to laugh now, silently. She put her hand out to hold his so that they could laugh together.

'Memsahib?'

A woman's voice. Minnie. She must have been standing well away from the louvred shutters. There was nothing to be seen of her head or her feet.

'It's all right Minnie. Quite all right, thank you. Go back to sleep and get some rest. I shall be back in bed very soon.'

She coughed, to underline her self possession. After a while she drank more of the brandy.

*

She drank more brandy. Straightened her body, leant back against the support of the raised lid, head against the wall, glanced at the empty throne beside her, then shut her eyes.

But when we went to parties, Tusker, just before we went in, you always took my arm. You helped me down from tongas and into tongas. Waiting on other people's verandahs for tongas, then, too, you took my arm, and in that way we waited. Arm in arm. Arm in arm. Throne by throne. What, now, Tusker? Urn by urn?

It's all right, Tusker. I really am not going to cry. I can't afford to cry. I have a performance to get through tomorrow. And another performance to get through on Wednesday. And on Thursday.

All I'm asking, Tusker, is did you mean it when you said I'd been a good woman to you? And if so, why did you leave me? Why did you leave me here? I am frightened to be alone, Tusker, although I know it is wrong and weak to be frightened—

—but now, until the end, I shall be alone, whatever I am doing, here as I feared, amid the alien corn, waking, sleeping, alone for ever and ever and I cannot bear it but mustn't cry and must must get over it but don't for the moment see how, so with my eyes shut, Tusker, I hold out my hand, and beg you, Tusker, beg, beg you to take it and take me with you. How can you not, Tusker? Oh, Tusker, Tusker, Tusker, how can you make me stay here by myself while you yourself go home?

PAUL SCOTT, *Staying On*, 1977

Queen Victoria's grief at the death of her husband is vividly conveyed in these pages from her journals and letters:

JOURNAL 13 December 1861

Found him very quiet and comfortably warm, and so dear and kind, called me '*gutes Fräuchen*' and kissed me so affectionately and so completely like himself, and I held his dear hands between mine . . . They gave him brandy every half-hour.

14 December 1861

Went over at 7 as I usually did. It was a bright morning; the sun just rising and shining brightly . . . Never can I forget how beautiful my darling looked lying there with his face lit up by the rising sun, his eyes unusually bright gazing as it were on unseen objects and not taking notice of me . . . I went out on the Terrace with Alice. The military band was playing at a distance and I burst out crying and came home again . . . I bent over him and said to him '*Es ist Kleines Fräuchen*' (it is your little wife) and he bowed his head; I asked him if he would give me '*ein Kuss*' (a kiss) and he did so. He seemed half dozing, quite quiet . . . I left the room for a moment and sat down on the floor in utter despair. Attempts at consolation from others only made me worse . . . Alice told me to come in . . . and I took his dear left hand which was already cold, though the breathing was quite gentle and I knelt down by him . . . Alice was on the other side, Bertie and Lenchen . . . kneeling at the foot of the bed . . . Two or three long but perfectly gentle breaths were drawn, the hand clasping mine and . . . all, all, was over . . . I stood up, kissed his dear heavenly forehead and called out in a bitter and agonising cry, 'Oh! my dear Darling!'

To Lord Derby 17 February 1862

The Queen wishes herself to express her thanks to Lord Derby [Leader of the Conservative Opposition] for the copy of his speech, and her satisfaction at his serving on the Committee of the Memorial [to the Prince Consort], so full of interest to her poor broken heart. She hopes to see him some day at Windsor, to which living grave she intends to return for a short while next week.

To express what the Queen's desolation and utter misery is, is almost impossible; every feeling seems swallowed up in that one of unbounded grief! She feels as though her life had ended on that dreadful day when she lost that bright Angel who was her idol, the life of her life; and time seems to have passed like one long, dark day!

She sees the trees budding, the days lengthen, the primroses coming out, but she thinks herself still in the month of December! The Queen toils away from morning till night, goes out twice a day, does all she is desired to do by her physician, but she wastes and pines, and there is that within her inmost soul, which seems to be undermining her existence!

To Palmerston 7 March 1863

The Queen is indeed most deeply touched and gratified at the extraordinary exhibition of loyalty and affection exhibited on the occasion of the arrival of our future daughter, a tribute which she well knows, and wishes all should know, is owing to her great and good husband, who led the Queen in the right path and to whom she owes everything and the country owes everything!

Four years after Albert's death, she was still unwilling to appear in public:

TO RUSSELL 22 January 1866

To enable the Queen to go through what she can only compare to an execution, it is of importance to keep the thought of it as much from her mind as possible, and therefore the going to Windsor to wait two whole days for this dreadful ordeal would do her positive harm.

The Queen has never till now mentioned this painful subject to Lord Russell, but she wishes once for all to just express her own feelings. She must, however, premise her observations by saying that she entirely absolves Lord Russell and his colleagues from any attempt ever to press upon her what is so very painful an effort. The Queen must say that she does feel very bitterly the want of feeling of those who ask the Queen to go to open Parliament. That the public should wish to see her she fully understands, and has no wish to prevent—quite the contrary; but why this wish should be of so unreasonable and unfeeling a nature, as to long to witness the spectacle of a poor, broken-hearted widow, nervous and shrinking, dragged in deep mourning, alone in State as a Show, where she used to go supported by her husband, to be gazed at, without delicacy of feeling, is a thing she cannot understand, and she never could wish her bitterest foe to be exposed to!

She will do it this time—as she promised it, but she owns she resents the unfeelingness of those who have clamoured for it. Of the suffering which it will cause her—nervous as she now is—she can give no idea, but she owns she hardly knows how she will go through it.

*

14 December 1899

Already thirty-eight years since that dreadful catastrophe which crushed and changed my life, and deprived me of my guardian angel, the best of husbands and most noble of men!

*

26 August 1900

This ever dear day has returned again without my beloved Albert being with me, who on this day, eighty-one years ago, came into the world as a blessing to so many, leaving an imperishable name behind him! How I remember the happy day it used to be, and preparing presents for him, which he would like! I thought much of the birthday spent at the dear lovely Rosenau in '45, when I so enjoyed being there, and where now his poor dear son, of whom he was so proud, has breathed his last.

CHRISTOPHER HIBBERT, ed., *Queen Victoria in her Letters and Journals. A Selection*, 1984

¶ *Queen Victoria died on 22 January 1901.*

My dearest dust, could not thy hasty day
Afford thy drowszy patience leave to stay
One hower longer: so that we might either
Sate up, or gone to bedd together?
But since thy finisht labour hath possest
Thy weary limbs with early rest,
Enjoy it sweetly: and thy widdowe bride
Shall soone repose her by thy slumbring side.
Whose business, now, is only to prepare
My nightly dress, and call to prayre:
Mine eyes wax heavy and ye day growes old.
The dew falls thick, my belovd growes cold.
Draw, draw ye closed curtaynes: and make roome:
My dear, my dearest dust; I come, I come.

LADY CATHERINE DYER, *from a monument erected by her in 1641 to her husband, Sir William Dyer*

THE WIDOWER

For a season there must be pain—
For a little, little space
I shall lose the sight of her face,
Take back the old life again
While She is at rest in her place.

For a season this pain must endure—
For a little, little while
I shall sigh more often than smile,
Till Time shall work me a cure,
And the pitiful days beguile.

For that season we must be apart.
For a little length of years,
Till my life's last hour nears,
And, above the beat of my heart,
I hear Her voice in my ears.

But I shall not understand—
Being set on some later love,
Shall not know her for whom I strove,
Till she reach me forth her hand,
Saying, 'Who but I have the right?'
And out of a troubled night
Shall draw me safe to the land.

RUDYARD KIPLING, 1890

ANDROMACHE'S LAMENT FOR HECTOR

'You've been torn from life,
my husband, in young manhood, and you leave me
empty in our hall. The boy's a child
whom you and I, poor souls, conceived; I doubt
he'll come to manhood. Long before, great Troy
will go down plundered, citadel and all,
now that you are lost, who guarded it
and kept it, and preserved its wives and children.
They will be shipped off in the murmuring hulls
one day, and I along with all the rest.
'You, my little one, either you come with me
to do some grinding labour, some base toil
for a harsh master, or an Achaean soldier
will grip you by the arm and hurl you down
from a tower here to a miserable death—
out of his anger for a brother, a father,

or even a son that Hector killed. Achaeans
in hundreds mouthed black dust under his blows.
He was no moderate man in war, your father,
and that is why they mourn him through the city.
Hector, you gave your parents grief and pain
but left me loneliest, and heart-broken.
You could not open your strong arms to me
from your deathbed, or say a thoughtful word,
for me to cherish all my life long
as I weep for you night and day.'

HOMER (*c.* 9th century BC), *The Iliad*

WIDOWER

Holes, spaces—not just in the small of the back
where her cushioned belly used to press,
but conversationally: missing even
those niggles about smoking less.

Missing most what was taken for granted—
the greatest, subtlest part, behind the eyes;
chasing a foam-sly figurehead without a ship,
hoping to check . . . the sea.

Holes, spaces, spawning themselves, endless.
That growths in her could leave such craters elsewhere!—
brutal vacuums sucking one to them
like bottles over boils, hurtful cure.

Sleepless, wanting the earthly full of her
even if varicose, unclean;
pacing, picking her ruins over
like a starved cat casing a blitzed town.

Frantic, straddling a lost smoulder
like a newspaper trying to nerve a flame;
giving up, piling the uncaught
with wet trash of regret, shame.

Holes, spaces: dragging a black tunnel,
her white shadow walking through one, still . . .
the space in the mind, cold air;
the hole in the heart, irreparable.

GEOFFREY HOLLOWAY

During those thirty minutes, he held the pencil over the sheet of paper and it moved and filled the pages in large letters. 'Your band of helpers are here, and welcome your visitor today. We are aware that he has suffered a recent bereavement, lost a person he loved. We can see that his heart is still very heavy and anguished. If we could help him, and others like him, to understand the nature of life and death, and relieve the pain at heart, we will have achieved our purpose. Death is only the vanishing point of the physical framework in which a personality is cast and functions; that same personality is unperceived before a conception, and will be lost sight of again at death, which we repeat is a vanishing point and not the end. . . .' Thus it went on sheet after sheet, at a pace of writing which was not normal—the pencil points broke off or tore through the paper. At one stage, the pencil said, 'The lady is here, but will not communicate with her husband directly yet. By and by, perhaps, when she is calmer. She is somewhat agitated today, since this is her first effort to communicate with her husband. She is disturbed by the grief of her husband. We on this side are directly affected by the thoughts emanating in your plane, and do our best to set your minds at peace. Today, she feels happy that there is an opening created and she could make some effort to influence your thoughts. . . .

'The lady wants to assure you that she exists but in a different state, she wants you to lighten your mind too, and not to let gloom weigh you down. She says now you are told I am here; by and by when you have attuned yourself, you will feel without proof or argument that I am at your side and that will transform your outlook. She advises you not to let anxiety develop about the child. She is well, and she will grow up well. I watch her.' . . .

Thirty minutes were over, and twenty-four hundred words had been written, which is an extraordinary speed of writing. . . .

Apart from the actual details of paper, pencil, and speed, I began to sense Rajam's presence at that table. What she is supposed to have said or Rao's pencil wrote was secondary. The actual presence felt at this sitting in the stillness and dimness of that little room had a profound effect on me. When I went home that evening, I felt lighter at heart. I remember there was a quarter moon in the sky—its light seemed deeper and more subtle than ever—the air seemed bracing—everything looked subtler and richer. . . .

Even if the whole thing was a grand fraud, it would not matter. What was important was the sensing of the presence in that room, which transformed my outlook. . . .

Apart from it all, what really mattered to me ultimately was the specific directions that she gave step by step in order to help me attain clarity of mind and receptivity. . . .

'Two nights ago, when you were about to fall asleep, your mind once again wandered off to the sick-bed scenes and the day you mourned my passing over. . . . No harm in your remembering those times, but at the root there is still a rawness and that interferes with your perception. Until you can think of me without pain, you will not succeed in your attempts. Train your mind properly and you will know that I am at your side.'

R. K. NARAYAN, *My Days*, 1975

AN ELEGY

Last, best-loved daughter of old Hsieh, you who foolishly ran off
with that penniless boy, who mended his clothes with patches
from your old clothes brought from home . . . and I teased you
for your gold hairpins, so we could trade for wine, and we drank
it with our dinners of berries and herbs picked cheap in the
fields, cooked over dry leaves from the fields . . . now, when
they pay me well, all I can give back to you is temple offerings.

Long long ago we could giggle at dying, but Death a magician
closed you in his hand and opened it suddenly empty. I have
locked your needlework away, I have given your clothes away . . .
my eyes are not strong enough: I am gentle, because you were,
to our serving-maids and men. Sometimes when I dream I dream
I shower you with gifts. All of us must know such sorrow . . . to
know it best you must first be poor and happy together.

Here I sit alone, here I sigh for both of us. How many beads must I still count upon my string of time? Better men than I have grown old without a son . . . a better poet sang to his dead wife who could not hear.

We never said that we would meet again in death. I have no hope beyond the darkness. All I have, is to stare into the night, seeing again and again that little worried wrinkle in your brow.

YUAN CHEN, *c.* AD 600

THE WIFE A-LOST

Since I noo mwore do zee your feäce,
Up steäirs or down below,
I'll zit me in the lwonesome pleäce,
Where flat-bough'd beech do grow;
Below the beeches' bough, my love,
Where you did never come,
An' I don't look to meet ye now,
As I do look at hwome.

Since you noo mwore be at my zide,
In walks in zummer het,
I'll goo alwone where mist do ride,
Drough trees a-drippen wet;
Below the raïn-wet bough, my love,
Where you did never come,
An' I don't grieve to miss ye now,
As I do grieve at hwome.

Since now bezide my dinner-bwoard
Your vaïce do never sound,
I'll eat the bit I can avvword,
A-yield upon the ground;
Below the darksome bough, my love,
Where you did never dine,
An' I don't grieve to miss ye now,
As I at hwome do pine.

Since I do miss your vaïce an' feäce
In praÿer at eventide,
I'll praÿ wi' woone sad vaïce vor greäce
To goo where you do bide;
Above the tree an' bough, my love,
Where you be gone avore,
An' be a-waïten vor me now,
To come vor evermwore.

WILLIAM BARNES, 1879

And no one ever told me about the laziness of grief. Except at my job—where the machine seems to run on much as usual—I loathe the slightest effort. Not only writing but even reading a letter is too much. Even shaving. What does it matter now whether my cheek is rough or smooth? They say an unhappy man wants distractions—something to take him out of himself. Only as a dog-tired man wants an extra blanket on a cold night; he'd rather lie there shivering than get up and find one. It's easy to see why the lonely become untidy; finally, dirty and disgusting.

*

It is incredible how much happiness, even how much gaiety, we sometimes had together after all hope was gone. How long, how tranquilly, how nourishingly, we talked together that last night!

And yet, not quite together. There's a limit to the 'one flesh'. You can't really share someone else's weakness, or fear or pain. What you feel may be bad. It might conceivably be as bad as what the other felt, though I should distrust anyone who claimed that it was. But it would still be quite different. When I speak of fear, I mean the merely animal fear, the recoil of the organism from its destruction; the smothery feeling; the sense of being a rat in a trap. It can't be transferred. The mind can sympathize; the body, less. In one way the bodies of lovers can do it least. All their love passages have trained them to have, not identical, but complementary, correlative, even opposite, feelings about one another.

We both knew this. I had my miseries, not hers; she had hers, not mine.

*

I have no photograph of her that's any good. I cannot even see her face distinctly in my imagination. Yet the odd face of some stranger seen in a crowd this morning may come before me in vivid perfection the moment I close my eyes tonight. No doubt, the explanation is simple enough. We have seen the faces of those we know best so variously, from so many angles, in so many lights, with so many expressions—waking, sleeping, laughing, crying, eating, talking, thinking—that all the impressions crowd into our memory together and cancel out in a mere blur. But her voice is still vivid. . . .

Already, less than a month after her death, I can feel the slow, insidious beginning of a process that will make the H. I think of into a more and more imaginary woman. . . .

The reality is no longer there to check me, to pull me up short, as the real H. so often did, so unexpectedly, by being so thoroughly herself and not me.

*

Slowly, quietly, like snow-flakes—like the small flakes that come when it is going to snow all night—little flakes of me, my impressions, my selections, are settling down on the image of her. The real shape will be quite hidden in the end. Ten minutes—ten seconds—of the real H. would correct all this. And yet, even if those ten seconds were allowed me, one second later the little flakes would begin to fall again. The rough, sharp, cleansing tang of her otherness is gone.

What pitiable cant to say 'She will live forever in my memory!' *Live?* That is exactly what she won't do. You might as well think like the old Egyptians that you can keep the dead by embalming them.

*

I can't settle down. I yawn, I fidget, I smoke too much. Up till this I always had too little time. Now there is nothing but time. Almost pure time, empty successiveness.

One flesh. Or, if you prefer, one ship. The starboard engine has gone. I, the port engine, must chug along somehow till we make harbour. Or rather, till the journey ends.

*

I think there is also a confusion. We don't really want grief, in its first agonies, to be prolonged: nobody could. But we want something else of which grief is a frequent symptom, and then we confuse the symptom with the thing itself. I wrote the other night that bereavement is not the truncation of married love but one of its regular phases—like the honeymoon. What we want is to live our marriage well and faithfully through that phase too. If it hurts (and it certainly will) we accept the pains as a necessary part of this phase. We don't want to escape them at the price of desertion or divorce. Killing the dead a second time. We were one flesh. Now that it has been cut in two, we don't want to pretend that it is whole and complete. We will be still married, still in love. Therefore we shall still ache. But we are not at all—if we understand ourselves—seeking the aches for their own sake. The less of them the better, so long as the marriage is preserved. And the more joy there can be in the marriage between dead and living, the better.

The better in every way. For, as I have discovered, passionate grief does not link us with the dead but cuts us off from them. This becomes clearer and clearer. It is just at those moments when I feel least sorrow—getting into my morning bath is usually one of them—that H. rushes upon my mind in her full reality, her otherness.

*

For in grief nothing 'stays put'. One keeps on emerging from a phase, but it always recurs. Round and round. Everything repeats. Am I going in circles, or dare I hope I am on a spiral?

But if a spiral, am I going up or down it?

*

Sorrow, however, turns out to be not a state but a process. It needs not a map but a history . . .

Grief is like a long valley, a winding valley where any bend may reveal a totally new landscape. As I've already noted, not every bend does. Sometimes the surprise is the opposite one; you are presented with exactly the same sort of country you thought you had left behind miles ago. That is when you wonder whether the valley isn't a circular trench. But it isn't. There are partial recurrences, but the sequence doesn't repeat.

*

It's not true that I'm always thinking of H. Work and conversation make that impossible. But the times when I'm not are perhaps my worst. For then, though I have forgotten the reason, there is spread over everything a vague sense of wrongness, of something amiss. Like in those dreams where nothing terrible occurs—nothing that would sound even remarkable if you told it at breakfast-time—but the atmosphere, the taste, of the whole thing is deadly. So with this. I see the rowan berries reddening and don't know for a moment why they, of all things, should be depressing. I hear a clock strike and some quality it always had before has gone out of the sound. What's wrong with the world to make it so flat, shabby, worn-out looking? Then I remember.

*

Something quite unexpected has happened. It came this morning early. For various reasons, not in themselves at all mysterious, my heart was lighter than it had been for many weeks. For one thing, I suppose I am recovering physically from a good deal of mere exhaustion. . . .

And suddenly at the very moment when, so far, I mourned H. least, I remembered her best. Indeed it was something (almost) better than memory; an instantaneous, unanswerable impression. To say it was like a meeting would be going too far. Yet there was that in it which tempts one to use those words. It was as if the lifting of the sorrow removed a barrier.

*

If, as I can't help suspecting, the dead also feel the pains of separation (and this may be one of their purgatorial sufferings), then for both lovers, and for all pairs of lovers without exception, bereavement is a universal and integral part of our experience of love. It follows marriage as normally as marriage follows courtship or as autumn follows summer. It is not a truncation of the process but one of its phases; not the interruption of the dance, but the next figure.

C. S. LEWIS, *A Grief Observed*, 1961

ON HIS DECEASED WIFE

Methought I saw my late espoused saint
Brought to me like Alcestis from the grave,
Whom Jove's great son to her glad husband gave,
Rescued from Death by force, though pale and faint.

Mine, as whom washed from spot of child-bed taint,
  Purification in the Old Law did save,
  And such as yet once more I trust to have
  Full sight of her in Heaven without restraint,
Came vested all in white, pure as her mind.
  Her face was veiled; yet to my fancied sight
  Love, sweetness, goodness, in her person shined
So clear as in no face with more delight.
  But, oh! as to embrace me she inclined,
  I waked, she fled, and day brought back my night.

JOHN MILTON, 1673

Are we both widowers within six months?——Marriage is a mystery even when, like mine, it is effected by a registrar (we neither of us could stand the Church of England service). I suppose every marriage is different: some are failures and never consummate themselves except physically, which means that they are not human marriages at all. I don't know why you did not live with Olive; her treating you badly (as I do) has nothing to do with it; all couples who live together treat each other badly occasionally, and treat third parties very well. But the relationship is unique and mysterious. I have known women with whom I could get on with less friction than with Charlotte; but their deaths have not affected me in the same way. Her ashes are preserved at Golders Green, and my instructions are that mine are to be inseparably mixed with them, after which they may be scattered to the winds or immured in Westminster Abbey for all I care. When I come across something intimate of her belongings I have a welling of emotion and quite automatically say something endearing to her. But I am not in the least desolate; on the contrary I am enjoying my solitude and have improved markedly in health since her death set me free. But you, who lived alone, feel desolate. There is no logic in it: it is a mystery.

GEORGE BERNARD SHAW, letter to Alfred Douglas, 29 February 1944

ON THE DEATH OF HIS WIFE

Since we were first married
Seventeen years have past.
Suddenly I looked up and she was gone.
She said she would never leave me.
My temples are turning white.
What have I to grow old for now?
At death we will be together in the tomb.
Now I am still alive,
And my tears flow without end.

*

IN BROAD DAYLIGHT I DREAM OF MY DEAD WIFE

Who says that the dead do not think of us?
Whenever I travel, she goes with me.
She was uneasy when I was on a journey.
She always wanted to accompany me.
While I dream, everything is as it used to be.
When I wake up, I am stabbed with sorrow.
The living are often parted and never meet again.
The dead are together as pure souls.

MEI YAO CH'EN, 11th century AD

EXEQUY TO HIS MATCHLESS NEVER-TO-BE-FORGOTTEN FRIEND

Accept, thou shrine of my dead saint,
Instead of dirges this complaint;
And for sweet flowers to crown thy hearse,
Receive a strew of weeping verse
From thy grieved friend, whom thou mightst see
Quite melted into tears for thee.
  Dear loss! since thy untimely fate
My task hath been to meditate
On thee, on thee! Thou art the book,
The library, whereon I look

Though almost blind. For thee, loved clay,
I languish out, not live, the day,
Using no other exercise
But what I practise with mine eyes.
By which wet glasses I find out
How lazily time creeps about
To one that mourns. This, only this,
My exercise and business is:
So I compute the weary hours
With sighs dissolvèd into showers.
  Nor wonder if my time go thus
Backward and most preposterous:
Thou hast benighted me. Thy set
This eve of blackness did beget,
Who wast my day (though overcast
Before thou hadst thy noon-tide past)
And I remember must in tears
Thou scarce hadst seen so many years
As day tells hours.

*

Sleep on, my Love, in thy cold bed
Never to be disquieted.
My last good night! Thou wilt not wake
Till I thy fate shall overtake:
Till age, or grief, or sickness must
Marry my body to that dust
It so much loves; and fill the room
My heart keeps empty in thy tomb.
Stay for me there: I will not fail
To meet thee in that hollow vale.
And think not much of my delay;
I am already on the way,
And follow thee with all the speed
Desire can make, or sorrows breed.
Each minute is a short degree
And every hour a step towards thee.
At night when I betake to rest,
Next morn I rise nearer my west
Of life, almost by eight hours sail

Than when sleep breathed his drowsy gale.
Thus from the sun my bottom steers,
And my day's compass downward bears.
Nor labour I to stem the tide
Through which to thee I swiftly glide.
'Tis true, with shame and grief I yield;
Thou, like the van, first took'st the field
And gotten hast the victory
In thus adventuring to die
Before me, whose more years might crave
A just precedence in the grave.
But hark! my pulse, like a soft drum,
Beats my approach, tells thee I come;
And slow howe'er my marches be
I shall at last sit down by thee.
The thought of this bids me go on
And wait my dissolution
With hope and comfort. Dear, (forgive
The crime) I am content to live
Divided, with but half a heart,
Till we shall meet and never part.

HENRY KING, to his wife Anne, who died in 1624.

12

The Marriage of True Minds

FINALLY, a celebration of marriage—idealized by poets, and also confirmed by couples who have been married for many years, yet write each other passionate love letters when separated. 'Love's not time's fool' . . .

Marriage is a School and Exercise of Virtue; and though Marriage hath Cares, yet the Single Life hath Desires, which are more troublesome and more dangerous, and often end in Sin; while the Cares are but Instances of Duty, and Exercises of Piety; and therefore if Single Life hath more Privacy of devotion, yet Marriage hath more Necessities and more variety of it, and is an Exercise of more Graces.

Marriage is the proper Scene of Piety and Patience, of the Duty of Parents and the Charity of Relations; here kindness is spread Abroad, and Love is united and made firm as a Centre; Marriage is the nursery of Heaven. The Virgin sends Prayers to God; but she carries but one soul to him: but the state of Marriage fills up the Number of the Elect, and hath in it the Labour of Love, and the Delicacies of Friendship, the Blessing of Society, and the Union of Hands and Hearts. It hath in it less of Beauty, but more of Safety than the Single Life; it hath more Care, but less Danger; it is more Merry, and more Sad; is fuller of Sorrows, and fuller of Joys: it lies under more Burdens, but is supported by all the Strengths of Love and Charity, and those Burdens are delightful.

JEREMY TAYLOR, *Sermon*, 1651

MARRIED LOVE

You and I
Have so much love,
That it
Burns like a fire,
In which we bake a lump of clay
Molded into a figure of you
And a figure of me.
Then we take both of them,
And break them into pieces,
And mix the pieces with water,
And mold again a figure of you,
And a figure of me.
I am in your clay.

You are in my clay.
In life we share a single quilt.
In death we will share one coffin.

KUAN TAO-SHENG, 13th century AD

THE BEST-BELOVED

E'en like two little bank-dividing brooks,
That wash the pebbles with their wanton streams,
And having ranged and searched a thousand nooks,
Meet both at length in silver-breasted Thames,
Where in a greater current they conjoin:
So I my best-beloved's am; so he is mine.

E'en so we met; and after long pursuit,
E'en so we joined, we both became entire;
No need for either to renew a suit,
For I was flax, and he was flames of fire
Our firm-united souls did more than twine;
So I my best-beloved's am; so he is mine . . .

Nor time, nor place, nor chance, nor death can bow
My least desires unto the least remove;
He's firmly mine by oath; I his by vow;
He's mine by faith; and I am his by love;
He's mine by water; I am his by wine:
Thus I my best-beloved's am; thus he is mine.

He is my altar; I, his holy place;
I am his guest; and he, my living food;
I'm his by penitence; he mine by grace;
I'm his by purchase; he is mine by blood;
He's my supporting elm; and I his vine:
Thus I my best-beloved's am; thus he is mine . . .

FRANCIS QUARLES, 1635

John Adams, who was to become the second President of the United States, and his wife Abigail were separated for the greater part of their married life, due to his many public appointments. Throughout, they kept up a passionate correspondence. Abigail always addressed her husband as her 'Friend', and often signed herself 'Portia'.

TO JOHN ADAMS.

16 October 1774.

My much loved Friend,

I dare not express to you, at three hundred miles' distance, how ardently I long for your return. I have some very miserly wishes, and cannot consent to your spending one hour in town, till, at least, I have had you twelve. The idea plays about my heart, unnerves my hand, whilst I write,—awakens all the tender sentiments, that years have increased and matured, and which, when with me, were every day dispensing to you. The whole collected stock of ten weeks' absence knows not how to brook any longer restraint, but will break forth and flow through my pen.

*

TO ABIGAIL ADAMS.

June 1777.

Next month completes three years that I have been devoted to the service of liberty. A slavery it has been to me, whatever the world may think of it. To a man whose attachments to his family are as strong as mine, absence alone from such a wife and such children would be a great sacrifice. But in addition to this separation what have I not done? What have I not suffered? What have I not hazarded? These are questions that I may ask you, but I will ask such questions of none else. Let the cymbals of popularity tinkle still. Let the butterflies of fame glitter with their wings. I shall envy neither their music nor their colors. The loss of property affects me little. All other hard things I despise, but the loss of your company and that of my dear babes for so long a time, I consider as a loss of so much solid happiness. The tender social feelings of my heart which have distressed me beyond all utterance in my most busy active scenes as well as in the numerous hours of melancholy solitude, are known only to God and my own soul.

*

TO JOHN ADAMS.

December 1782.

'If you had known,' said a person to me the other day, 'that Mr Adams would have remained so long abroad, would you have consented that he should have gone?' I recollected myself a moment, and then spoke the real dictates of my heart. 'If I had known, Sir, that Mr Adams could have effected what he has done, I would not only have submitted to the absence I have endured, painful as it has been, but I would not have opposed it, even though three years more should be added to the number, (which Heaven avert!) I feel a pleasure in being able to sacrifice my selfish passions to the general good, and in imitating the example, which has taught me to consider myself and family but as the small dust of the balance, when compared with the great community.' . . .

Ever remember me, as I do you, with all the tenderness, which it is possible for one object to feel for another, which no time can obliterate, no distance alter, but which is always the same in the bosom of

Portia.

TO MY DEAR AND LOVING HUSBAND

If ever two were one, then surely we.
If ever man were loved by wife, then thee;
If ever wife was happy in a man,
Compare with me ye women if you can.
I prize thy love more than whole mines of gold,
Or all the riches that the East doth hold.
My love is such that rivers cannot quench,
Nor ought but love from thee, give recompence.
Thy love is such I can no way repay,
The heavens reward thee manifold I pray.
Then while we live, in love lets so persever,
That when we live no more, we may live ever.

ANNE BRADSTREET, 1678

William Wordsworth married his childhood friend Mary Hutchinson in 1802. In June 1810, shortly after the birth of their fifth child, William set off with his sister Dorothy on a leisurely journey to visit old friends. During the few weeks of their separation, William and Mary wrote each other numerous letters, interspersing family news and reports on children and harvests with declarations of their love for each other.

(*William to Mary*): 11 August 1810

Every day every hour every moment makes me feel more deeply how blessed we are in each other, how purely how faithfully how ardently, and how tenderly we love each other; I put this last word last because, though I am persuaded that a deep affection is not uncommon in married life, yet I am confident that a lively, gushing, thought-employing, spirit-stirring, passion of love, is very rare even among good people. . . . We have been parted my sweet Mary too long, but we have not been parted in vain, for wherever I go I am admonished how blessed, and almost peculiar a lot mine is. . . .

O Mary I love you with a passion of love which grows till I tremble to think of its strength; your children and the care which they require must fortunately steal between you and the solitude and the longings of absence—when I am moving about in travelling I am less unhappy than when stationary, but then I am at every moment, I will not say reminded of you, for you never I think are out of my mind 3 minutes together however I am engaged, but I am every moment seized with a longing wish that you might see the objects which interest me as I pass along, and not having you at my side my pleasure is so imperfect that after a short look I had rather not see the objects at all.

Two years later, despite their resolution never to part again, William went to London to make his peace with Coleridge.

(*Mary to William*): 23 May 1812

Yet I *do* not regret that this separation has been, for it is worth no small sacrifice to be thus assured, that instead of weakening, our union has strengthened—a hundred fold strengthened those yearnings towards each other which I used so strongly to feel at Gallows Hill—& in which you sympathized with me at that time—that these feelings are mutual now, I have the fullest proof, from thy letters & from their power & the power of absence over my whole frame—Oh

William I can not tell thee how I love thee, & thou must not desire it—but feel it, O feel it in the fullness of thy soul & *believe* that I am the happiest of Wives & of Mothers & of all Women the most blessed—and, if it be gratitude to acknowledge this, not by words, but by actions—by supposing that this *must* be understood because it is so—that the spirit feels & it *must* communicate then William, I am the most grateful—not only to thee but to every thing that breathes & to the Great God the giver of all good—But I must stop or I know not whither I shall be carried & instead of composing myself by retiring I shall unfit myself for receiving the Party from Hereford whom we are expecting—

A letter from Karl Marx to his wife Jenny, when she was visiting her dying mother in Trier. They had been married for thirteen years at this time:

21 June 1856

My heart's beloved

I am writing you again, because I am alone and because it troubles me always to have a dialogue with you in my head, without your knowing anything about it or hearing it or being able to answer. Poor as your photograph is, it does perform a service for me, and I now understand how even the 'Black Madonna' [ikon], the most disgraceful portrait of the Mother of God, could find indestructible admirers, indeed even more admirers than the good portraits. In any case, those Black Madonna pictures have never been more kissed, looked at, and adored than your photograph, which, although not black, is morose, and absolutely does not reflect your darling, sweet, kissable *dolce* face. But I improve upon the sun's rays, which have painted falsely, and find that my eyes, so spoiled by lamplight and tobacco, can still paint, not only in dream but also while awake. I have you vivaciously before me, and I carry you on my hands, and I kiss you from head to foot, and I fall on my knees before you, and I groan 'Madame, I love you'. And I truly love you more than the Moor of Venice ever loved. . . .

Momentary absence is good, for in constant presence things seem too much alike to be differentiated. Proximity dwarfs even towers, while the petty and commonplace, at close view, grow too big. Small habits, which may physically irritate and take on emotional form, disappear when the immediate object is removed from the eye. Great passions, which through proximity assume the form of petty routine, grow and again take on their natural dimension on account of the magic of distance. So it is with my love. You have only to be snatched away from me even in a mere dream, and I know immediately that the time has only served, as do sun and rain for plants, for growth. The moment you are absent, my love for you shows itself to be what it is, a giant, in which are crowded together all the energy of my spirit and all the character of my heart.

There are actually many females in the world, and some among them are beautiful. But where could I find again a face, whose every feature, even every wrinkle, is a reminder of the greatest and sweetest memories of my life? Even my endless pains, my irreplaceable losses I read in your sweet countenance, and I kiss away the pain when I kiss your sweet face. 'Buried in her arms, awakened by her kisses'—namely, in your arms and by your kisses, and I grant the Brahmins and Pythagoras their doctrine of regeneration and Christianity its doctrine of resurrection. . . . Goodbye, my sweet heart. I kiss you and the children many thousand times.

Yours,
Karl.

John Jay Chapman (1862–1933) was a New York lawyer, social reformer, and writer. He fell in love with Minna Timmins, but had no idea she reciprocated his feelings. Thinking that someone was making her unhappy, he thrashed the offender, but was then so conscience-stricken that he plunged his hand into a fire, and burned it so badly that the hand had to be amputated. Minna then wrote him letters 'of passionate devotion', and they were married in 1889. He wrote her the following letter while on a business trip.

I have sealed up each one of these letters thinking I had done—and then a wave of happiness has come over me—remembering you—only you, my Minna—and the joy of life. Where were you, since the beginning of the world? But now you are here, about me in every space, room, sunlight, with your heart and arms and the light of your soul—and the strong vigor your presence. It was not a waste desert in Colorado. It is not a waste time, for you are here and many lives packed into one life, and the green shoot out of the heart of the plant, springing up blossoms in the night, and many old things have put on immortality and lost things have come back knocking within, from before the time I was conceived in the womb, there were you also. And what shall we say of the pain! it was false—and the rending, it was necessary. It was the breaking down of the dams that ought not to have been put up—but being up it was the sweeping away of them that the waters might flow together.

This is a love letter, is it not? How long is it since I have written you a love letter, my love, my Minna? Was the spring hidden that now comes bubbling up overflowing curb and coping-stone, washing my feet and my knees and my whole self? How are the waters of the world sweet—if we should die, we have drunk them. If we should sin—or separate—if we should fail or secede—we have tasted of happiness—we must be written in the book of the blessed. We have had what life could give, we have eaten of the tree of knowledge, we have known—we have been the mystery of the universe.

Is love a hand or a foot—is it a picture or a poem or a fireside—is it a compact or a permission or eagles that meet in the clouds—No, no, no, no. It is light and heat and hand and foot and ego. If I take the wings of the morning and remain in the uttermost parts of the sea, there art thou also—He descended into Hell and on the third day rose again—and there art thou also—in the lust or business—in the stumbling and dry places, in sickness and health—every sort of sickness there also—what matter is it what else the world contains—if you only are in it in every part of it? I can find no corner of it without you—my eyes would not see it. It is empty—I have seen all that is there and it is nothing, and over creation are your wings.

Have we not lived three years now together—and daily nearer—grafted till the very sap of existence flows and circulates between us—till I know as I write this—your thoughts—till I know as a feeling, a hope, a thought, passes through me—it is yours? Why the agony of those old expressions and attempts to come by diligent, nervous, steady, fixing of the eye on the graver's tool, as if the prize depended on drawing it straight, those pounds of paper and nights of passionate composition—did they indeed so well do their work that the goal was carried—or was it the silent communion—of the night—even after days of littleness or quarrel that knitted us together? It does not matter, love, which it was. It put your soul so into my body that I don't speak to you to convey meaning. I write only for joy and happiness. How diligently have we set fact to fact and consideration against consideration during the past years—as if we were playing dominoes for our life. How cloudy I have been—dragging you down, often nailing useless nails cutting up and dissecting, labeling, crucifying small things—and there was our great love over us, growing, spreading—I wonder we do not shine—or speak with every gesture and accent giving messages from the infinite—like a Sibyl of Michael Angelo. I wonder people do not look after us in the street as if they had seen an angel.

Tuo Giovanni.

¶ *Minna died in 1897 after the birth of her third child.*

You are my only love. You have me completely in your power. I *know* and *feel* that if I am to write anything fine and noble in the future I shall do so only by listening at the doors of your heart. I would like to go through life side by side with you, telling you more and more until we grew to be one being together until the hour should come for us to die.

JAMES JOYCE, letter to Nora, 1909

EPITAPH UPON A YOUNG MARRIED COUPLE, DEAD AND BURIED TOGETHER

To these, whom Death again did wed,
This grave's their second marriage-bed.
For though the hand of fate could force
'Twixt soul and body a divorce,
It could not sunder man and wife,
'Cause they both livèd but one life.
Peace, good Reader. Do not weep.
Peace, the lovers are asleep.
They, sweet turtles, folded lie
In the last knot love could tie;
And though they lie as they were dead,
Their pillow stone, their sheets of lead,
(Pillow hard, and sheets not warm)
Love made the bed; they'll take no harm.
Let them sleep: let them sleep on,
Till this stormy night be gone,

Till the eternal morrow dawn;
Then the curtains will be drawn
And they wake into a light
Whose day shall never die in night.

RICHARD CRASHAW, 1652

I reopen my envelope to tell you I have recd your dear letter of the 28th. I reciprocate intensely the feelings of love & devotion you show to me. My greatest good fortune in a life of brilliant experience has been to find you, & to lead my life with you. I don't feel far away from you out here at all. I feel vy near in my heart; & also I feel that the nearer I get to honour, the nearer I am to you.

WINSTON CHURCHILL, letter to his wife, 1915

A DEDICATION TO MY WIFE

To whom I owe the leaping delight
That quickens my senses in our wakingtime
And the rhythm that governs the repose of our sleepingtime,
The breathing in unison

Of lovers whose bodies smell of each other
Who think the same thoughts without need of speech
And babble the same speech without need of meaning.

No peevish winter wind shall chill
No sullen tropic sun shall wither
The roses in the rose-garden which is ours and ours only

But this dedication is for others to read:
These are private words addressed to you in public.

T. S. ELIOT, 1959

To darling Carlotta, my wife, who for twenty-three years has endured with love and understanding my rotten nerves, my lack of stability, my cussedness in general . . .

I am old and would be sick of life, were it not that you, Sweetheart, are here, as deep and understanding in your love as ever—and I as deep in my love for you as when we stood in Paris, Premier Arrondissement, on July 22, 1929, and both said faintly 'Oui'!

¶ *A dedication written on the flyleaf of* Moon for the Misbegotten *by Eugene O'Neill in July 1952, a few months before his death. Carlotta was his third wife, and the marriage had survived through many quarrels and estrangements.*

Marriage is one long conversation, chequered by disputes. The disputes are valueless; they but ingrain the difference; the heroic heart of woman prompting her at once to nail her colours to the mast. But in the intervals, almost unconsciously, and with no desire to shine, the whole material of life is turned over and over, ideas are struck out and shared, the two persons more and more adapt their notions one to suit the other, and in process of time, without sound of trumpet, they conduct each other into new worlds of thought.

R. L. STEVENSON, 1881

SONG

Why should your face so please me
That if one little line should stray
Bewilderment would seize me
And drag me down the tortuous way
Out of the noon into the night?
But so, into this tranquil light
You raise me.

How could our minds so marry
That, separate, blunder to and fro,
Make for a point, miscarry,
And blind as headstrong horses go?
Though now they in their promised land
At pleasure travel hand in hand
Or tarry.

This concord is an answer
To questions far beyond our mind
Whose image is a dancer.
All effort is to ease refined
Here, weight is light; this is the dove
Of love and peace, not heartless love
The lancer.

And yet I still must wonder
That such an armistice can be
And life roll by in thunder
To leave this calm with you and me.
This tranquil voice of silence, yes,
This single song of two, this is
A wonder.

EDWIN MUIR, 1946

THE TRANCE

Sometimes, apart in sleep, by chance,
You fall out of my arms, alone,
Into the chaos of your separate trance.
My eyes gaze through your forehead, through the bone,
And see where in your sleep distress has torn
Its violent path, which on your lips is shown
And on your hands and in your dream forlorn.

Restless, you turn to me, and press
Those timid words against my ear
Which thunder at my heart like stones.
'Mercy,' you plead, Then 'Who can bless?'
You ask. 'I am pursued by Time,' you moan.
I watch that precipice of fear
You tread, naked in naked distress.

To that deep care we are committed
Beneath the wildness of our flesh
And shuddering horror of our dream,
Where unmasked agony is permitted.

Our bodies, stripped of clothes that seem,
And our souls, stripped of beauty's mesh,
Meet their true selves, their charms outwitted.

This pure trance is the oracle
That speaks no language but the heart,
Our angel with our devil meets
In the atrocious dark nor do they part
But each each forgives and greets,
And their mutual terrors heal
Within our married miracle.

STEPHEN SPENDER, 1941–9

THE FROZEN BOY

Ah, let your arms about me in the bed
Tighten in love, and wordless joys
Force from you words no counterfeit could shed!
Within that grasp, that whispered noise,
You catch two lovers—two, I said—
You catch two lovers by the heart:
One known to you, your man now, tame or wild,
One prisoned long ago by time, a child.

The man you know. As for the boy you hold,
He stands with frozen-postured stare—
At four, unwanted. Hammered at by cold,
Anger, and guilt, in his despair
He makes No-saying batten hold
On him; No-saying courts less grief.
His piteous strategy!—What loneliness
It brought us both!—Love, free him with our Yes'.

Don't fear that frozen boy as I have feared
Him down the years. A secret son—
A father—he has spectrally appeared
To stay my hand, to chide my fun,
To stifle hope. Yet hope is reared:
So may that ice-faced boy depart—

Dismiss the waif with blessings, and outlive
His long-drawn No with our affirmative.

BRIAN W. ALDISS, 1982

MIDCENTURY LOVE LETTER

Stay near me. Speak my name. Oh, do not wander
By a thought's span, heart's impulse, from the light
We kindle here. You are my sole defender
(As I am yours) in this precipitous night,
Which over earth, till common landmarks alter,
Is falling, without stars, and bitter cold.
We two have but our burning selves for shelter.
Huddle against me. Give me your hand to hold.

So might two climbers lost in mountain weather
On a high slope and taken by the storm,
Desperate in the darkness, cling together
Under one cloak and breathe each other warm.
Stay near me. Spirit, perishable as bone,
In no such winter can survive alone.

PHYLLIS McGINLEY, 1961

REPRISE

Geniuses of countless nations
Have told their love for generations
Till all their memorable phrases
Are common as goldenrod or daisies.
Their girls have glimmered like the moon,
Or shimmered like a summer noon,
Stood like lily, fled like fawn,
Now the sunset, now the dawn,
Here the princess in the tower
There the sweet forbidden flower.
Darling, when I look at you
Every aged phrase is new,
And there are moments when it seems
I've married one of Shakespeare's dreams.

OGDEN NASH, 1949

How do I love thee? Let me count the ways.
I love thee to the depth and breadth and height
My soul can reach, when feeling out of sight
For the ends of Being and ideal Grace.
I love thee to the level of every day's
Most quiet need, by sun and candle-light.
I love thee freely, as men strive for right;
I love thee purely, as they turn from praise.
I love thee with the passion put to use
In my old griefs, and with my childhood's faith.
I love thee with a love I seemed to lose
With my lost saints—I love thee with the breath,
Smiles, tears, of all my life!—and, if God choose,
I shall but love thee better after death.

ELIZABETH BARRETT BROWNING, 1850

A Letter written to her husband, the Russian poet Osip Mandelstam, by Nadezhda Mandelstam, in October 1938, when he had been arrested not long after returning from a three-year exile. In January the following year she learned that he was dead. 'Millions of women', she says, 'wrote such letters—to their husbands, sons, brothers or simply to sweethearts' in Stalinist Russia.

Osia, my beloved, faraway sweetheart!
I have no words, my darling, to write this letter that you may never read, perhaps. I am writing it into empty space. Perhaps you will come back and not find me here. Then this will be all you have left to remember me by.

Osia, what a joy it was living together like children—all our squabbles and arguments, the games we played, and our love. Now I do not even look at the sky. If I see a cloud, who can I show it to?

Remember the way we brought back provisions to make our poor feasts in all the places where we pitched our tent like nomads? Remember the good taste of bread when we got it by a miracle and ate it together? And our last winter in Voronezh. Our happy poverty, and the poetry you wrote. I remember the time we were coming back once from the baths, when we bought some eggs or sausage, and a cart went by loaded with hay. It was still cold and I was freezing in my short jacket (but nothing like what we must suffer now: I know how cold you are). That day comes back to me now. I understand so clearly, and ache from the pain of it, that those winter days with all their troubles were the greatest and last happiness to be granted us in life.

My every thought is about you. My every tear and every smile is for you. I bless every day and every hour of our bitter life together, my sweetheart, my companion, my blind guide in life.

Like two blind puppies, we were, nuzzling each other and feeling so good together. And how fevered your poor head was, and how madly we frittered away the days of our life. What joy it was, and how we always knew what joy it was.

Life can last so long. How hard and long for each of us to die alone. Can this fate be for us who are inseparable? Puppies and children, did we deserve this? Did you deserve this, my angel? Everything goes on as before. I know nothing. Yet I know everything—each day and hour of your life are plain and clear to me as in a delirium.

You came to me every night in my sleep, and I kept asking what had happened, but you did not reply.

In my last dream I was buying food for you in a filthy hotel restaurant. The people with me were total strangers. When I had bought it, I realized I did not know where to take it, because I do not know where you are.

When I woke up, I said to Shura: 'Osia is dead.' I do not know whether you are still alive, but from the time of that dream, I have lost track of you. I do not know where you are. Will you hear me? Do you know how much I love you? I could never tell you how much I love you. I cannot tell you even now. I speak only to you, only to you. You are with me always, and I who was such a wild and angry one and never learned to weep simple tears—now I weep and weep and weep.

It's me: Nadia. Where are you?

Farewell.

Nadia.

Let me not to the marriage of true minds
Admit impediments. Love is not love
Which alters when it alteration finds,
Or bends with the remover to remove:
O, no! it is an ever-fixed mark,
That looks on tempests and is never shaken;

It is the star to every wandering bark,
Whose worth's unknown, although his height be taken.
Love's not Time's fool, though rosy lips and cheeks
Within his bending sickle's compass come;
Love alters not with his brief hours and weeks,
But bears it out even to the edge of doom.
  If this be error, and upon me prov'd,
  I never writ, nor no man ever lov'd.

SHAKESPEARE, 1609

ACKNOWLEDGEMENTS

THE editor and publisher are grateful for permission to include the following copyright material in this anthology.

All rights in respect of the Authorized King James Version of the Holy Bible are vested in the Crown in the United Kingdom and controlled by Royal Letters Patent.

Excerpts from the Revised Standard Version of the Bible, copyright 1946, 1952, 1971 by the Division of Christian Education of the National Council of the Churches of Christ in the USA, used by permission.

Dannie Abse, 'In my fashion', © Dannie Abse 1981, first published by Hutchinson in *Way Out of the Centre*. Reprinted by permission of Anthony Sheil Associates on the author's behalf.

Edward F. Albee, from *Who's Afraid of Virginia Woolf?*, © 1962 by Edward Albee. Reprinted by permission of Jonathan Cape Ltd. on the author's behalf and Atheneum Publishers, an imprint of Macmillan Publishing Company.

Edward Albee, from *A Delicate Balance* (1967).

Brian W. Aldiss, 'The Frozen Boy' from *Farewell to a Child* (Priapus Press, 1982). Reprinted by permission of the author.

A. Alvarez, from *Life After Marriage: People in Divorce*. Reprinted by permission of Macmillan, London and Basingstoke.

Yehuda Amichai, 'A Pity: We Were Such a Good Invention' from *Selected Poems*, trans. Assia Gotmann. Published in the United States in *Selected Poetry of Yehuda Amichai* eds. Chana Bloch and Stephen Mitchell, © 1986 by Chana Bloch and Stephen Mitchell. Reprinted by permission of Jonathan Cape Ltd. and Harper & Row, Publishers, Inc.

Daisy Ashford, from *The Young Visiters*. Reprinted by permission of Chatto & Windus on behalf of the Estate of Daisy Ashford.

Ausonius, 'To his wife' from *Medieval Latin Lyrics*, trans. Helen Waddell. Reprinted by permission of Constable Publishers.

John Berryman, excerpt from *The Freedom of the Poet*, copyright 1940, 1944, 1945, 1947, 1948, 1949, © 1956, 1959, 1961, 1964, 1966 by John Berryman, copyright renewed © 1972 by John Berryman; copyright 1951, 1953, © 1960, 1965, 1975, 1976 by Kate Berryman, renewed © 1973, 1975, 1976 by Kate Berryman. Reprinted by permission of Farrar, Straus & Giroux, Inc.

Alan Bold, trans., *see under* Lorca.

James Cameron, from *Point of Departure*. Reprinted by permission of David Higham Associates Ltd.

G. V. Catullus, 'Guardian of Helicon, Urania's son . . .', trans. Frederic Raphael and Kenneth McLeish. Reprinted by permission of A. P. Watt Ltd., on behalf of Volatic Ltd. and Kenneth McLeish.

Geoffrey Chaucer, extract from 'The Wife of Bath's Prologue' and from 'The Merchant's Tale' from *The Canterbury Tales* by Geoffrey Chaucer, trans. Nevill Coghill (Penguin Classics, Revised Edition, 1977), copyright 1951 by Nevill Coghill, © Nevill Coghill, 1958, 1960, 1975, 1977. Reprinted by permission of Penguin Books Ltd.

Winston S. Churchill, from a letter to his wife, Nov. 1909, from *Winston S. Churchill* by Randolph S. Churchill, Volume II, Companion Part 2, 1909–11, © 1969 C. & T. Publications Ltd. All rights reserved; from letters to his wife, 17 July 1915, and December 1915, from *Winston S. Churchill* by Martin Gilbert, Vol. III, Part 2, May 1915–Dec.1915, © 1973 C. & T. Publications Ltd. Reprinted by permission of William Heinemann Ltd., and Houghton Mifflin Company.

Ivy Compton-Burnett, from *Daughters and Sons*. Copyright I. Compton-Burnett 1937. Reprinted by permission of Curtis Brown Ltd., London.
Evan S. Connell, from *Mrs Bridges and Mr Bridges* (1959) [Viking].
E. E. Cummings, 'may i feel said he' from *Complete Poems 1913–1962*. Published in the United States in *No Thanks*, ed. George James Firmage, copyright 1935 by E. E. Cummings, © 1968 by Marion Morehouse Cummings, © 1973, 1978 by the Trustees for the E. E. Cummings Trust, © 1973, 1978 by George James Firmage. Reprinted by permission of Grafton Books, a division of William Collins & Son and Liveright Publishing Corporation.
Joan Didion, excerpt from 'Marrying Absurd', from *Slouching Towards Bethlehem*, © 1967, 1968 by Joan Didion. Reprinted by permission of Farrar, Straus & Giroux, Inc. and Wallace Literary Agency Inc.
Douglas Dunn, 'France' from *Elegies*. Reprinted by permission of Faber & Faber Ltd.
T. S. Eliot, extracts from *The Cocktail Party*, copyright 1950 by T. S. Eliot and renewed 1978 by Esme Valerie Eliot; 'A Dedication to my Wife' from *Collected Poems 1909–1962*, copyright 1936 by Harcourt Brace Jovanovich, Inc., © 1963, 1964 by T. S. Eliot. Reprinted by permission of Faber & Faber Ltd., Harcourt Brace Jovanovich, Inc., and Farrar, Straus & Giraux Ltd.
Alice Thomas Ellis, from *The Other Side of The Fire* (1983). Reprinted by permission of Gerald Duckworth & Co. Ltd.
Hans Magnus Enzensberger, 'Die Scheidung' ('The Divorce') from *Die Furie des Verschwindens*, © Suhrkamp Verlag Frankfurt am Main 1980. First published in English *The Rialto*, Summer 1985, trans. Rheinhold Grimm and Felix Pollak. Translation © 1985 by Reinhold Grimm and Felix Pollak. Reprinted by permission of Suhrkamp Verlag, Professor Dr Reinhold Grimm and Mrs Pollak.
Nora Ephron, from *Heartburn*, © 1983 by Nora Ephron. Reprinted by permission of Alfred A. Knopf, Inc. and International Creative Management.
Euripides, lines 228–51 from *The Medea*, trans. Rex Warner, 1944, in Grene and Lattimore (eds.), *The Complete Greek Tragedies*, Volume III, copyright 1942, 1952, 1955, 1956, 1958, 1960 by the University of Chicago. Reprinted by permission of the University of Chicago Press.
Simon Fallowfield, letter of proposal, reprinted by permission of Steve Race.
from *Fellbrigg: The Story of a House* by R. W. Ketton-Cremer, published by Century Hutchinson.
Shulamith Firestone, from *The Dialectics of Sex: The Case for Feminist Revolution* copyright © 1970. Reprinted by permission of Laurence Pollinger Ltd., and William Morrow & Co.
Gustave Flaubert, from *Madame Bovary*, trans. Alan Russell (Penguin Classics, 1950), copyright 1950 by Alan Russell. Reprinted by permission of Penguin Books Ltd.
Eleanor Flexner, from *Century of Struggle: The Woman's Rights Movement in the United States*, © 1959 by the President and Fellows of Harvard College. Reprinted with permission.
Ford Madox Ford, from *The Good Soldiers*. Reprinted by permission of The Bodley Head on behalf of the Estate of Ford Madox Ford.
E. M. Forster, from *Howard's End* copyright 1921 by E. M. Forster. Reprinted by permission of Edward Arnold and Alfred A. Knopf, Inc.
Sigmund Freud, from *New Introductory Lectures*, in *The Standard Edition of the Complete Psychological Works of Sigmund Freud*, trans. and ed. by James Strachey. Reprinted by permission of Sigmund Freud Copyrights Ltd., The Institute of Psycho-Analysis and The Hogarth Press, and the US publisher, W. W. Norton & Company, Inc.
Betty Friedan, from *The Feminine Mystique* (1952). Reprinted by permission of W. W.

Norton & Company, Inc. and Victor Gollancz Ltd.
Erich Fromm, from *The Art of Loving*. Reprinted by permission of Unwin Hyman Ltd.
Stella Gibbons, from *Cold Comfort Farm* (Longman, 1933). Reprinted by permission of David Higham Associates.
Kahlil Gibran, 'Love one another . . .' and 'Your children are not your children . . .' both reprinted from *The Prophet* by Kahlil Gibran, by permission of Alfred A. Knopf Inc. Copyright 1923 by Kahlil Gibran and renewed 1951 by Administrators C. T. A. of Kahlil Gibran Estate, and Mary G. Gibran.
Johann Wolfgang von Goethe, from *Elective Affinities* trans. R. J. Hollingdale (Penguin Classics, 1971), translation copyright R. J. Hollingdale, 1971. Reprinted by permission of Penguin Books Ltd.
Robert Graves, 'With her lips only' and 'Call it a good morning' from *Collected Poems 1975*, © 1975 by Robert Graves. Reprinted by permission of A. P. Watt Ltd. on behalf of The Executors of the Estate of Robert Graves, and Oxford University Press Inc.
Germaine Greer, from *The Female Eunuch*. Reprinted by permission of Grafton Books, a division of William Collins Sons & Co Ltd.
Edna Healey, from *Wives of Fame*. Reprinted by permission of Sidgwick & Jackson Ltd.
Geoffrey Holloway, 'Widower' from *All I Can Say* by Geoffrey Holloway (Anvil Press Poetry, 1978). Reprinted by permission of the author.
Homer, from 'Andromache's Lament for Hector' from *The Iliad*, trans. Robert Fitzgerald, translation © 1974 by Robert Fitzgerald. Used by permission of Doubleday, a division of Bantam, Doubleday, Dell Publishing Group, Inc.
Henrik Ibsen, excerpt from *The Doll's House* from *Four Major Plays* (1987), trans. and ed. James McFarlane. Reprinted by permission of Oxford University Press.
from *John Jay Chapman and His Letters*, ed. M. A. deWolfe Howe, copyright 1937 by M. A. DeWolfe Howe, © renewed 1965 by Quincy Howe, Helen Howe Allen, Mark DeWolfe Howe. Reprinted by permission of Houghton Mifflin Company.
from *The Journals of Dorothy Wordsworth*, ed. Mary Moorman (1971). Reprinted by permission of Oxford University Press.
James Joyce, from *Ulysses*. Copyright 1914, 1918 by Margaret Caroline Anderson and renewed 1942, 1946 by Nora Joseph Joyce. Reprinted by permission of Random House Inc., and The Bodley Head on behalf of the Executors of the James Joyce Estate.
C. J. Jung, from *Collected Works*, Vol. 17, trans. F. C. Hull. Reprinted by permission of Princeton University Press and Routledge.
Franz Kafka, from *The Diaries of Franz Kafka, 1910–1923*, ed. Max Brod, trans. Jospeh Kresh. Copyright 1948 and renewed 1976 by Shocken Books Inc. Published in the UK by Martin Secker & Warburg Ltd. Reprinted by permission of the publishers.
Rudyard Kipling, 'The Widower' from *Rudyard Kipling's Verse: Definitive Edition*. Reprinted by permission of Doubleday, a division of Bantam, Doubleday, Dell Publishing Group, Inc.
Sheila Kitzinger, from *The Experience of Childbirth*. Reprinted by permission of Victor Gollancz Ltd. and the Taplinger Publishing Co.
Kuan Tao-Sheng, 'Married Love' from *The Seasons of Women*, by Penelope Washbourn, © 1979 by Penelope Washbourn. Reprinted by permission of Harper & Row, Publishers, Inc.
Milan Kundera, from *The Unbearable Lightness of Being*, trans. Michael Henry Heim, © 1984 by Harper & Row, Publishers, Inc. Reprinted by permission of Faber & Faber Ltd. and Harper & Row, Publishers, Inc.

Choderlos de Lachos, from *Les Liaisons Dangereuses*, trans. P. W. K. Stone (Penguin Classics, 1961), © P. W. K. Stone, 1961. Reprinted by permission of Penguin Books Ltd.

R. D. Laing, 'If she did not love you . . .' from *Sonnets*, © 1979 R. D. Laing, published by Michael Joseph Ltd. Used by permission of the publisher.

Philip Larkin, extract from 'The Whitsun Weddings'; 'Talking in Bed' and 'An Arundel Tomb', all from *The Whitsun Weddings*. Reprinted by permission of Faber & Faber Ltd.

D. H. Lawrence, from *Women in Love*, copyright 1920, 1922 by David Herbert Lawrence, copyright 1948, 1950 by Frieda Lawrence; from *Kangaroo*, copyright 1923 by Thomas Seltzer, Inc., copyright 1951 by Frieda Lawrence; from *The Rainbow*, copyright 1915 by David Herbert Lawrence, copyright 1943 by Frieda Lawrence. Reprinted by permission of Viking Penguin, a division of Penguin Books USA, Inc.

Laurie Lee, from *Cider with Rosie*. Reprinted by permission of The Hogarth Press on behalf of the author.

from *The Letters of Anton Pavlovitch Tchekov to Olga Leonardovna Knipper*, trans. Constance Garnett (George H. Doran Co., New York, 1924).

Alun Lewis, 'Goodbye' from *Ha Ha Amongst The Trumpets* (Allen & Unwin, 1945). Reprinted by permission of Unwin Hyman Ltd.

C. S. Lewis, from *A Grief Observed*. Reprinted by permission of Faber & Faber Ltd. (published in the United States by Seabury Press).

Maurice Lindsay, 'Love's Anniversaries (for Joyce)' from *Collected Poems* (Paul Harris, Edinburgh 1979). Reprinted by permission of the author.

Frederico Garcia Lorca, 'The Unfaithful Wife' reprinted from *The Picador Book of Erotic Verse*, translation © 1978 Alan Bold and used with his permission.

Alison Lurie, from *War Between the Tates*, © 1974 by Alison Lurie. Reprinted by permission of William Heinemann Ltd. and Random House, Inc.

Lu Yu, 'Idleness' from *One Hundred Poems from the Chinese*, trans. Kenneth Rexroth, copyright © 1971 by Kenneth Rexroth. Reprinted by permission of New Directions Publishing Corp.

Nadezhda Mandelstam, from *Hope Abandoned*, trans. from the Russian by Max Hayward. © 1972 by Atheneum Publishers. Reprinted by permission of Collins Publishers and Atheneum Publishers, an imprint of Macmillan Publishing Company.

Martial, Epigram 104, Book XI, 'Wife, there are some points', reprinted from *The Picador Book of Erotic Verse* (1983), ed. Alan Bold.

John Masters, from *The Road Past Mandalay*, © 1961 by John Masters, © 1972, © renewed 1989 by Barbara Masters. Reprinted by permission of Michael Joseph Ltd. and Brandt & Brandt Literary Agents Inc.

François Mauriac, from *The Knot of Vipers*, trans. G. Hopkins (Eyre & Spottiswoode, 1957). Used by permission.

Mary McCarthy, from *The Group* (Weidenfeld, 1954). Reprinted by permission of A. M. Heath & Co. Ltd. on behalf of the author.

Carson McCullers, from *The Member of the Wedding* (Barrie & Jenkins 1946). Reprinted by permission of the publisher on behalf of the Estate of Carson McCullers.

Ian McEwan, from *The Child in Time* (1987). Reprinted by permission of Jonathan Cape Ltd. on behalf of the author (Rogers Coleridge & White).

Phyllis McGinley, 'Daniel at Breakfast', 'The 5:32' and 'Midcentury Love Letter' from *Times Three*, copyright 1941, 1950, 1953, 1954, renewed © 1969 by Phyllis McGinley, renewed © 1978 by Julie Hayden and Phyllis Hayden Blake, renewed © 1981, 1982 by Viking Penguin. All rights reserved. Reprinted by permission of

Martin Secker & Warburg Ltd. and Viking Penguin, a division of Penguin Books USA, Inc.

Ved Mehta, from *The Ledge Between the Streams*, © 1982, 1983, 1984 by Ved Mehta. Reprinted by permission of Collins Publishers and W. W. Norton and Company Inc.

Mei Yao Ch'en, 'On the death of his wife' and 'In Broad Daylight I Dream of My Dead Wife' from *One Hundred Poems from the Chinese*, trans. Kenneth Rexroth, copyright © Kenneth Rexroth. Reprinted by permission of New Directions Publishing Corp.

Mary Midgley, from *Beast and Man*, published by Cornell University Press.

Nancy Mitford, extract from the inside cover of her Appointments Diary, 1941. Reprinted by permission of the Peters Fraser & Dunlop Group Ltd.

John Mortimer, from *Clinging to the Wreckage*, © 1982 by Advanpress Ltd. Reprinted by permission of Ticknor & Fields, a Houghton Mifflin Company, and Weidenfeld & Nicolson Ltd.

Wolfgang Amadeus Mozart, excerpts from *The Letters of Mozart and His Family*, trans. and ed. Emily Anderson. Reprinted by permission of Macmillan, London and Basingstoke.

Edwin Muir, 'The Confirmation' and 'Song' from *Collected Poems* by Edwin Muir, © 1960 by Willa Muir. Reprinted by permission of Oxford University Press, Inc. and Faber & Faber Ltd.

Iris Murdoch, from *Nuns and Soldiers*, © Iris Murdoch, 1980. All rights reserved. Reprinted by permission of Chatto & Windus Ltd., on the author's behalf and Viking Penguin.

R. K. Narayan, from *My Days*. © by R. K. Narayan, published by Chatto & Windus, 1975. Reprinted by permission of the publisher and Anthony Sheil Associates Ltd. on the author's behalf.

Ogden Nash, 'Reprise', from *Versus*, copyright 1947 by Ogden Nash. First published in *The New Yorker*. 'The Strange Case of Mr Ormantude's Bride', from *Verses from 1929 on*, copyright 1942 by Ogden Nash. Reprinted by permission of Little, Brown and Company; 'Did Someone Say Babies?' from *Free Wheeling*, copyright 1931 by Ogden Nash and 'The Third Jungle Book' from *I Wouldn't Have Missed It*, published in the US in *Good Intentions*, copyright 1942 by Ogden Nash. Reprinted by permission of Andre Deutsch Ltd. and Little, Brown and Company.

A. S. Neill, from *Summerhill*, © 1960 by A. S. Neill. Reprinted by permission of Curtis Brown Ltd. and Victor Gollancz Ltd.

Friedrich Nietzsche, from *Thus Spake Zarathustra*, trans. R. J. Hollingdale (Penguin Classics, rev. edn., 1969), © R. J. Hollingdale, 1961, 1969. Reprinted by permission of Penguin Books Ltd.

Tillie Olsen, from *Tell Me A Riddle*, © Tillie Olsen 1962. Reprinted by permission of the author.

Eugene O'Neill, from *The Iceman Cometh*. Copyright © Yale University. Reprinted by permission of Jonathan Cape Ltd. on behalf of the Executors of the Eugene O'Neill Estate; 'Inscription to Carlotta' on the MS of *The Moon for the Misbegotten*, published in *Inscriptions: Eugene O'Neill to Carlotta Monterey O'Neill*, New Haven: Yale University Library, 1960. Used by Permission of the Yale Committee on Literary Property.

John Osborne, from *Look Back In Anger*, © 1957 by John Osborne, © renewed 1985. Reprinted by permission of Faber & Faber Ltd. and Harold Ober Associates Inc.

Ovid, from *Metamorphoses* VIII, 613 ff., adapted by Kenneth McLeish, from *Myths and Legends of Ancient Rome* (Longman, 1987). Reprinted by permission of A. P. Watt Ltd., on behalf of Kenneth McLeish.

Samuel Pepys, from *The Diary of Samuel Pepys*, ed. R. Latham and W. Matthews

(Unwin Hyman Ltd.). Reprinted by permission of the Peters Fraser & Dunlop Group Ltd.

Plato, from *The Symposium*, trans. Benjamin Jowett. Used by permission of The Jowett Copyright Trustees.

from *Queen Victoria In Her Letters and Journals*, ed. and sel. Christopher Hibbert, © 1984 Christopher Hibbert. All rights reserved. Reprinted by permission of John Murray (Publishers) Ltd. and Viking Penguin, a division of Penguin Books USA, Inc.

Piers Paul Read, from *A Married Man* (1979) Reprinted by permission of Martin Secker & Warburg Ltd.

Peter Redgrove, 'Far Star' from *In the Hall of the Saurian* (1987). Reprinted by permission of Martin Secker & Warburg Ltd.

Bertrand Russell, from 'Love, An Escape from Loneliness', reprinted by permission of Anton Felton.

George Sand, excerpt from a letter to her half-brother, Hippolyte Chatiron, from *George Sand, In Her Own Words*, ed. Joseph Barry (1974). Reprinted by permission of Doubleday, a division of Bantam, Doubleday, Dell Publishing Group, Inc.

Paul Scott, from *Staying On*, © Paul Scott 1977. Reprinted by permission of William Heinemann Ltd. and David Higham Associates Ltd.

Bernard Shaw, from *Man and Superman* (1903); letter to Alfred Douglas, 1944 from *Bernard Shaw and Alfred Douglas, a Correspondence*, ed. Mary Hyde (John Murray, 1982) and 'Divorce, in fact, is not the destruction of marriage . . .', all reprinted by permission of The Society of Authors on behalf of the Bernard Shaw Estate.

A. C. Robin Skynner, from *One Flesh, Separate Persons*, © 1976 by A. C. Robin Skynner. Reprinted by permission of Constable Publishers and Curtis Brown Ltd., London.

Stevie Smith, 'Autumn' from *The Collected Poems of Stevie Smith* (Penguin Modern Classics/New Directions), © 1972 by Stevie Smith. Reprinted by permission of James MacGibbon, Literary Executor and New Directions Publishing Corp.; 'Marriage I Think' from *Me Again: The Uncollected Writings of Stevie Smith*. All Stevie Smith writings and drawings © James MacGibbon 1937, 1938, 1942, 1950, 1957, 1962, 1966, 1971, 1978. Published by Virago Press Ltd. 1981. Reprinted by permission of Virago Press.

from *The Spectator*, vol. iv, ed. Donald Bond (1965). Reprinted by permission of Oxford University Press.

Stephen Spender, 'The Trance', copyright 1947 by Stephen Spender, reprinted from *Collected Poems* by permission of Faber & Faber Ltd. and Random House, Inc.

Anne Stevenson, 'The Marriage', extract from 'Correspondences' and 'A Love Letter: Ruth Arbeiter to Major Paul Maxwell', from *Selected Poems 1956–1986* (1987). Reprinted by permission of Oxford University Press.

'Sukey, you shall be my wife' from *The Oxford Nursery Rhyme Book* assembled by Iona and Peter Opie (1955). Reprinted by permission of Oxford University Press.

Rosemary Sutcliff, from *Blue Remembered Hills* © 1983. Reprinted by permission of The Bodley Head on the author's behalf and William Morrow & Co.

David Sutton, 'Not to be Born' from *Absences and Celebrations*. Reprinted by permission of Chatto & Windus on behalf of the author.

Takahishi Shinkichi, 'Birth', from *The Penguin Book of Japanese Verse* trans. Geoffrey Bownas and Anthony Thwaite (Penguin Books, 1964), © Geoffrey Bownas and Anthony Thwaite, 1964. Reprinted by permission of Penguin Books Ltd.

A. S. J. Tessimond, 'Heaven' from *The Collected Poems of A. S. J. Tessimond*, ed. Hubert Nicholson (Whiteknights Press, Reading). Used by permission of Hubert Nicholson.

Dylan Thomas, from a letter to his wife Caitlin from *Collected Lettere*, ed. Paul Ferris

(Dent, 1985) published in US in *Selected Letters of Dylan Thomas*, © 1965, 1966 by the Trustees for the Copyrights of Dylan Thomas; from *Under Milk Wood* (London: Dent/New York: New Directions, 1954), copyright 1954 by New Directions Publishing Corporation. All Rights Reserved; 'On the Marriage of a Virgin' from *The Poems of Dylan Thomas*, copyright 1943 by New Directions Publishing Corporation. Reprinted by permission of David Higham Associates Ltd. and New Directions Publishing Corporation.

Flora Thompson, from *Lark Rise to Candleford* (1939). Reprinted by permission of Oxford University Press.

James Thurber, 'The Wooing of Mr Monroe' published in the UK in *Vintage Thurber* and reprinted by permission of Hamish Hamilton Ltd. Published in the US by Harper & Row in *The Owl in the Attic*, copyright 1931, © 1959 James Thurber, reprinted by permission of Lucy Kroll Agency.

Anthony Thwaite, 'Marriages' from *Poems 1953–1988* by Anthony Thwaite (Hutchinson, 1989): Anthony Thwaite © 1977, 1989. Used by permission of the author.

J. R. R. Tolkien, from a letter to his son Michael, 6–8 March 1941, from *Letters of J. R. R. Tolkien*, eds. H. Carpenter and Christopher Tolkien (Allen & Unwin, 1981). Reprinted by permission of Unwin Hyman Ltd.

L. N. Tolstoy, from *Anna Karenin* by L. N. Tolstoy, trans. Rosemary Edmonds (Penguin Classics, 1954), © Rosemary Edmonds, 1954; from War and Peace, trans. Rosemary Edmonds (Penguin Classics, rev. edn., 1978), © Rosemary Edmonds, 1957, 1978. Reprinted by permission of Penguin Books Ltd.

Sophia Tolstoy, from *The Diaries of Sophia Tolstoy*, trans. Cathy Porter © 1985 by Cathy Porter. Reprinted by permission of Random House, Inc. and Jonathan Cape Ltd.

Arthur Waley, 'My Lord Summons Me' from *Chinese Poems* (1946), trans. Arthur Waley. Reprinted by permission of Unwin Hyman Ltd.

Francis Warner, 'Epithalamium' from *Collected Poems 1960–1984* (Colin Smythe, 1985). Reprinted by permission of the publisher.

Evelyn Waugh, from *The Letters of Evelyn Waugh*, ed. Mark Amory, © 1980 the Estate of Laura Waugh, © 1980 in the Introduction and compilation by Mark Amory. Reprinted by permission of Weidenfeld (Publishers) Ltd. and Houghton Mifflin Company.

Fay Weldon, from *The Life and Loves of a She-Devil*, © 1983 by Fay Weldon. Reprinted by permission of Hodder & Stoughton Ltd. and Pantheon Books, a Division of Random House, Inc.

H. G. Wells, from *An Englishman Looks at the World* (Cassell, 1914), and from *Love and Mr Lewisham* (1900). Reprinted by permission of A. P. Watt Ltd., on behalf of The Literary Executors of the Estate of H. G. Wells.

Rebecca West, from *Cousin Rosamund*, © The Estate of Rebecca West, 1985. Published by Virago Press Ltd. 1984. Reprinted by permission of Virago Press.

Patrick White, from *The Tree of Man* (Penguin, 1961). Reprinted by permission of Barbara Mobbs for the author.

P. G. Wodehouse, from *Jeeves in the Offiing* (Herbert Jenkins, 1960) and from *The Adventures of Sally* (Herbert Jenkins, 1922). Reprinted by permission of Century and A. P. Watt Ltd., on behalf of The Trustees of the Wodehouse Trust No. 3.

Virginia Woolf, from *To The Lighthouse*, copyright 1927 by Harcourt Brace Jovanovich, Inc. and renewed 1955 by Leonard Woolf, reprinted by permission of Harcourt Brace Jovanovich, Inc., and Chatto & Windus on behalf of the Executors of the Virginia Woolf Estate.

There are instances where we have been unable to trace or contact the copyright holder before our printing deadline. We apologize for this apparent negligence. If notified the publisher will be pleased to rectify any errors or omissions at the earliest opportunity.

INDEX OF AUTHORS